Studies in Computational Intelligence

Volume 672

Series editor

Janusz Kacprzyk, Polish Academy of Sciences, Warsaw, Poland
e-mail: kacprzyk@ibspan.waw.pl

About this Series

The series "Studies in Computational Intelligence" (SCI) publishes new developments and advances in the various areas of computational intelligence—quickly and with a high quality. The intent is to cover the theory, applications, and design methods of computational intelligence, as embedded in the fields of engineering, computer science, physics and life sciences, as well as the methodologies behind them. The series contains monographs, lecture notes and edited volumes in computational intelligence spanning the areas of neural networks, connectionist systems, genetic algorithms, evolutionary computation, artificial intelligence, cellular automata, self-organizing systems, soft computing, fuzzy systems, and hybrid intelligent systems. Of particular value to both the contributors and the readership are the short publication timeframe and the worldwide distribution, which enable both wide and rapid dissemination of research output.

More information about this series at http://www.springer.com/series/7092

Huimin Lu · Yujie Li
Editors

Artificial Intelligence and Computer Vision

Editors
Huimin Lu
Department of Electrical Engineering and Electronics
Kyushu Institute of Technology
Fukuoka
Japan

Yujie Li
School of Information Engineering
Yangzhou University
Yangzhou
China

ISSN 1860-949X ISSN 1860-9503 (electronic)
Studies in Computational Intelligence
ISBN 978-3-319-83477-1 ISBN 978-3-319-46245-5 (eBook)
DOI 10.1007/978-3-319-46245-5

Printed on acid-free paper

This Springer imprint is published by Springer Nature
The registered company is Springer International Publishing AG
The registered company address is: Gewerbestrasse 11, 6330 Cham, Switzerland

Preface

The integration of artificial intelligence and computer vision technologies has become a topic of increasing interest for both researchers and developers from academic fields and industries worldwide. It is foreseeable that artificial intelligence will be the main approach of the next generation of computer vision research. The explosive number of artificial intelligence algorithms and increasing computational power of computers have significantly extended the number of potential applications for computer vision. It has also brought new challenges to the vision community. The aim of this book is to provide a platform to share up-to-date scientific achievements in this field. The papers were chosen based on review scores submitted by the members of the program committee and underwent further rigorous rounds of review.

In Computer Vision for Ocean Observing, Huimin Lu, Yujie Li and Seiichi Serikawa present the application of computer vision technologies for ocean observing. This chapter also analyzes the recent trends of ocean exploration approaches.

In Fault Diagnosis and Classification of Mine Motor Based on RS and SVM, Xianmin Ma, Xing Zhang and Zhanshe Yang propose a fault diagnosis method for the mine hoist machine fault diversity and redundancy of fault data based on rough sets and support vector machine.

In Particle Swarm Optimization Based Image Enhancement of Visual Cryptography Shares, Mary Shanthi Rani M. and Germine Mary G. propose a particle swarm optimization based image enhancement of visual cryptography shares. The proposed algorithm guarantees highly safe, secure, quick and quality transmission of secret image with no mathematical operation needed to reveal the secret.

In Fast Level Set Algorithm for Extraction and Evaluation of Weld Defects in Radiographic Images, Boutiche Y. proposes a fast level set algorithm for extraction and evaluation of weld defects in radiographic images. The segmentation is assured using a powerful implicit active contour implemented via fast algorithm. The curve is represented implicitly via binary level set function. Weld defect features are computed from the segmentation result.

In Efficient Combination of Color, Texture and Shape Descriptor, Using SLIC Segmentation for Image Retrieval, N. Chifa, A. Badri, Y. Ruichek, A. Sahel and K. Safi present a novel method of extraction and combination descriptor to represent image. First we extract a descriptor shape (HOG) from entire image, and second we apply the proposed method for segmentation, and then we extract the color and texture descriptor from each segment.

In DEPO: Detecting Events of Public Opinion in Microblog, Guozhong Dong, Wu Yang and Wei Wang propose DEPO, a system for detecting events of public opinion in microblog. In DEPO, abnormal messages detection algorithm is used to detect abnormal messages in the real-time microblog message stream. Combined with events of public opinion (EPO) features, each abnormal message can be formalized as EPO features using microblog-oriented keywords extraction method.

In Hybrid Cuckoo Search Based Evolutionary Vector Quantization for Image Compression, Karri Chhiranjeevi and Uma Ranjan Jena propose a hybrid cuckoo search (HCS) algorithm that optimizes the LBG codebook with less convergence time by taking McCulloch's algorithm based levy flight distribution function and variant of searching parameters.

In Edge and Fuzzy Transform Based Image Compression Algorithm: edgeFuzzy , Deepak Gambhir and Navin Rajpal propose an edge-based image compression algorithm in fuzzy transform (F-transform) domain. Input image blocks are classified either as low-intensity blocks, medium-intensity blocks or high-intensity blocks depending on the edge image obtained using the Canny edge detection algorithm. Based on the intensity values, these blocks are compressed using F-transform. Huffman coding is then performed on compressed blocks to achieve reduced bit rate.

In Real-Time Implementation of Human Action Recognition System Based on Motion Analysis, Kamal Sehairi, Cherrad Benbouchama, Kobzili El Houari, and Chouireb Fatima propose a pixel streams-based FPGA implementation of a real-time system that can detect and recognize human activity using Handel-C.

In Cross-Modal Learning with Images, Texts and Their Semantics, Xing Xu proposes a novel model for cross-modal retrieval problem. The results well demonstrate the effectiveness and reasonableness of the proposed method.

In Light Field Vision for Artificial Intelligence, Yichao Xu and Miu-ling Lam review the recent process in light field vision. The newly developed light field vision technique shows a big advantage over conventional computer vision techniques.

It is our sincere hope that this volume provides stimulation and inspiration, and that it will be used as a foundation for works to come.

Fukuoka, Japan
Yangzhou, China
August 2016

Huimin Lu
Yujie Li

Contents

Contributors

A. Badri Faculty of Sciences and Techniques (FSTM), EEA&TI Laboratory, Hassan II University of Casablanca, Mohammedia, Morocco

Cherrad Benbouchama Laboratoire LMR, École Militaire Polytechnique, Algiers, Algeria

Yamina Boutiche Research Center in Industrial Technologies CRTI, ex CSC., Algiers, Algeria; Faculté des Sciences de L'ingénieur, Département d' Electronique, Université Saad Dahlab de Blida, Blida, Algeria

N. Chifa Faculty of Sciences and Techniques (FSTM), EEA&TI Laboratory, Hassan II University of Casablanca, Mohammedia, Morocco

Karri Chiranjeevi Department of Electronics and Tele-Communication Engineering, Veer Surendra Sai University of Technology, Burla, Odisha, India

Fatima Chouireb Laboratoire LTSS, Université Amar Telidji Laghouat, Laghouat, Algeria

Guozhong Dong Information Security Research Center, Harbin Engineering University, Harbin, China

Deepak Gambhir School of Information and Communication Technology, Guru Gobind Singh Inderprastha University, Dwarka, New Delhi, India

Umaranjan Jena Department of Electronics and Tele-Communication Engineering, Veer Surendra Sai University of Technology, Burla, Odisha, India

El Houari Kobzili Laboratoire LMR, École Militaire Polytechnique, Algiers, Algeria

Miu-ling Lam School of Creative Media, City University of Hong Kong, Kowloon Tong, Hong Kong

Yujie Li School of Information Engineering, Yangzhou University, Yangzhou, China

Huimin Lu Department of Electrical and Electronic Engineering, Kyushu Institute of Technology, Kitakyushu, Japan; Institute of Oceanology, Chinese Academy of Sciences, Qingdao, China; State Key Laboratory of Ocean Engineering, Shanghai Jiaotong University, Shanghai, China; ARC Robotic Vision Centre, Queensland University of Technology, Brisbane, Australia

Xianmin Ma College of Electrical and Control Engineering, Xi'an University of Science & Technology, Shaanxi, China

G. Germine Mary Department of Computer Science, Fatima College, Madurai, Tamil Nadu, India

P.M.K. Prasad Department of Electronics and Communication Engineering, GMR Institute of Technology, Rajam, India

Navin Rajpal School of Information and Communication Technology, Guru Gobind Singh Inderprastha University, Dwarka, New Delhi, India

M. Mary Shanthi Rani Department of Computer Science and Applications, Gandhigram Rural Institute—Deemed University, Dindigul, Tamil Nadu, India

Y. Ruichek IRIES-SET-UTBM, Belfort Cedex, France

K. Safi Faculty of Sciences and Techniques (FSTM), EEA&TI Laboratory, Hassan II University of Casablanca, Mohammedia, Morocco

A. Sahel Faculty of Sciences and Techniques (FSTM), EEA&TI Laboratory, Hassan II University of Casablanca, Mohammedia, Morocco

Kamal Sehairi Laboratoire LTSS, Université Amar Telidji Laghouat, Laghouat, Algeria

Seiichi Serikawa Department of Electrical and Electronic Engineering, Kyushu Institute of Technology, Kitakyushu, Japan

Wei Wang Information Security Research Center, Harbin Engineering University, Harbin, China

Yichao Xu School of Creative Media, City University of Hong Kong, Kowloon Tong, Hong Kong

Xing Xu School of Computer Science and Engineering, University of Electronic Science and Technology of China, Chengdu, China

Wu Yang Information Security Research Center, Harbin Engineering University, Harbin, China

Zhanshe Yang College of Electrical and Control Engineering, Xi'an University of Science & Technology, Shaanxi, China

Xing Zhang College of Electrical and Control Engineering, Xi'an University of Science & Technology, Shaanxi, China

Computer Vision for Ocean Observing

Huimin Lu, Yujie Li and Seiichi Serikawa

Abstract There have been increased developments in ocean exploration using autonomous underwater vehicles (AUVs) and unmanned underwater vehicles (UUVs). However, the contrast of underwater images is still a major issue for application. It is difficult to acquire clear underwater images around underwater vehicles. Since the 1960s, sonar sensors have been extensively used to detect and recognize objects in oceans. Due to the principles of acoustic imaging, sonar-imaged images have many shortcomings, such as a low signal to noise ratio and a low resolution. Consequently, vision sensors must be used for short-range identification because sonars yield to low-quality images. This thesis will concentrate solely on the optical imaging sensors for ocean observing. Although the underwater optical imaging technology makes a great progress, the recognition of underwater objects also remains a major issue in recent days. Different from the common images, underwater images suffer from poor visibility due to the medium

H. Lu (✉) · S. Serikawa
Department of Electrical and Electronic Engineering, Kyushu Institute of Technology, Kitakyushu 8048550, Japan
e-mail: luhuimin@ieee.org

H. Lu
Institute of Oceanology, Chinese Academy of Sciences, Qingdao 266071, China

H. Lu
State Key Laboratory of Ocean Engineering, Shanghai Jiaotong University, Shanghai 200240, China

H. Lu
ARC Robotic Vision Centre, Queensland University of Technology, Brisbane 3000, Australia

Y. Li
School of Information Engineering, Yangzhou University, Yangzhou 225127, China
e-mail: liyujie@yzu.edu.cn

H. Lu and Y. Li (eds.), *Artificial Intelligence and Computer Vision*,
Studies in Computational Intelligence 672, DOI 10.1007/978-3-319-46245-5_1

scattering and light distortion. First of all, capturing images underwater are difficult, mostly due to attenuation caused by light. The random attenuation of the light mainly causes the haze appearance along with the part of the light scattered back from the water. In particular, the objects at a distance of more than 10 m are almost indistinguishable because of absorption. Furthermore, when the artificial light is employed, it can cause a distinctive footprint on the seafloor. In this paper, we will analysis the recent trends of ocean exploration approaches.

Keywords Ocean observing · Computer vision

1 Introduction

Ocean observations [1] are being developed and deployed by scientists, researchers and institutions around the world oceans for monitoring the status of ocean. Some observatories are cabled, For example, the Ocean Networks Canada Observatory [2], contains VENUS and NEPTUNE Canada cabled networks. It enables real-time interactive experiments, for measuring ocean health, ecosystems, resources, natural hazards and marine conservation. Some observatories are moored or made up of surface buoys, such as NOAA Ocean Climate Observation System [3]. The observations near the equator are of particular important to climate. Besides of monitoring the air-water exchange of heat and water, the moored buoys provide platforms for instrumentation to measure the air-water exchange of carbon dioxide in the tropics. Some observatories are remote sensed, such as Japanese Ocean Flux Data Sets with Use of Remote Sensing Observation [4]. It is used for monitoring the changes of heat, water and momentum with atmosphere at ocean surface.

Interestingly, there are some excellent systems for ocean observing, such as Global Ocean Observing System proposed by Henry. Stommel WHOI [5]. More than 30 countries are joined in this program. However, until now this system also has some drawbacks. First, the system is not fully built-out because of funding issues. Second, most of subsystems are not at full operational capacity. Many of them are funded through research programs rather than operational. Third, deep ocean (under 2000 m) is very under-sampled-issue of technology and cost.

In this thesis, we firstly review the ocean observation systems in technology level. Then, analysis the feasibility of recent systems and propose some novel technologies for improving recent optical imaging systems.

1.1 Remote Sensing

Costal and ocean resources are fully affected by ocean, land-atmospheric physics, biology and geology. Extreme events and environmental disasters require the satellite remote sensing to track currents, map ocean productivity, assess winds and waves. The development and usage of satellites that complement ship-based observations, moored and other autonomous sensors and models. It will provide high quality data more frequently, allowing for improving site-specific forecasts of weather, water conditions, and resource distribution [6].

Fortunately, there are existed some ocean remote sensing systems in the world. The Global Ocean Observing System (GOOS), the Global Climate Observing System (GCOS), and the Global Earth Observing System of Systems (GEOSS) are coordinated with U.S. plans for satellite remote sensing.

1.2 *In Situ Sensing*

In 1985, the Data Buoy Cooperation Panel (DBLP) was estimated, as a joint body of the World Meteorological Organization (WMO) and Intergovernmental Oceanographic Commission (IOC) of UNESCO. The DBLP aims to share international coordination and assist those providing and using observations from data buoys, within the meteorological and oceanographic communities [7].

Until now, DBLP has 1250 global surface drifting buoy array with 5 degrees resolution. It also has 120 moorings for combining the global tropical moored buoy network.

1.3 *Underwater Sensing*

Ocean bottom sensor nodes plan an important role for underwater sensing. They are used for monitoring the oceanographic data collection, pollution monitoring, offshore exploration, disaster prevention, assisted navigation and tactical surveillance. Unmanned underwater vehicles (UUVs) and Autonomous underwater vehicles (AUVs) are equipped with underwater sensors. And most of them are used to find application in exploration of natural underwater resources and gathering of scientific data in collaborative monitoring missions.

Recently, wireless underwater acoustic networking is the enabling technology for these applications. Although underwater communication has been studied since WWII, there are also some disadvantages. Firstly, real time monitoring is impossible. Secondly, control of the underwater systems is difficult. Thirdly, it is limited to record the amount of data during the monitoring mission [8]. Many researchers are currently focusing on developing next generalization network for terrestrial wireless ad hoc and sensor networks.

2 Underwater Imaging Systems

2.1 *Acoustic Imaging*

Sound can be used to make a map of reflected intensities, which is called sonogram. These sonar images are often resemble optical images, and the level of details higher than the traditional ways. However, if the deployed forms of environment, the sonogram can be completely confused, and it takes significant experience before it can be interpreted correctly [9].

Sonograms are created by the devices which emit beam-formed pulses toward the seafloor. The sonar beams are narrow in one direction and wide in the other direction, emitted down from the transducer to the objects. The intensities of the acoustic reflection from the seafloor are called "fan-shaped", which likes an image. As the beam is moved, the reflections will depict a series of image lines perpendicular to the direction of motion. When stitched together "along track", the lines will produced a single image of the seabed or objects [9].

It is necessary that the movement can be achieved by rotation of the transducer array, just like in sector scanning sonars for remotely operated vehicle (ROV), where they are used as navigational aids, such as conventional ship-radars. However, the array is towed on a cable behind the ship, and because of the lines imaged are perpendicular to the length axis of ship, this equipment is known as side-scan sonar.

In the 1970s, the long range GLORIA side-scan sonar was developed. It was used to survey the large oceanic areas, operated at relatively low frequencies (6 kHz) and was used to produce images of continental shelves world-wide [10]. Recently, the 30 kHz Towed Ocean Bottom Instrument (TOBI) multisensor is used instead of it. To reach a higher resolution of the sonar image, it is possible to either increase the frequency or to increase the number of elements of the transducer array [11]. On the other hand, signal processing techniques are used for improving its performance.

About 50 years ago, there have been a lot of people who attempt to design an acoustic camera. The first successful set was the EWATS system, which was created in the 1970s and had 200 lines of resolution and maximum of 10 m range. In the 2000s, DIDSON [12], Echoscope [13], BlueView [14] or the other acoustic camera are designed for serving the underwater.

While the above mentioned acoustic imaging cameras perform well, they also have the challenges in the measurement of the seafloor or objects. In order to assess and survey small-scale bed morphology features in ocean, coastal, river, the issue of increase the accuracy and resolution of imaging sonars is also remaining. Another challenge is to reduce the cost of multi-beam sonar, so as to facilitate a wider application of the technique.

2.2 Optical Imaging

Optical imaging sensors can provide much information updated at high speed and they are commonly used in many terrestrial and air robotic application. However, because of the interaction between electromagnetic waves and water, optical imaging systems and vision systems need to be specifically designed to be able to use in underwater environment [15].

Underwater images have specific characteristics that should be taken into account during the gathering process and processing process. Light attenuation, scattering, non-uniform lighting, shadows, color filtering, suspended particles or abundance of marine life on top or surrounding the target of interest are frequently found in typical underwater scenes [16].

One effect of the inherent optical properties of ocean is that it becomes darker and darker with the deepening water depth. As the water depth increases, the sunlight from the sun is absorbed and scattered. For example, in the relatively clean ocean water, the euphotic depth is 200 m or less [17]. In addition, the spectral composition of sunlight also changes with the water depth. Absorption is greater for long wavelengths than for short, this is prominent effect even at shallow depth with 10 m. Therefore most underwater images taken in natural light (sunlight) will appear blue or green on videos and thus for all but the most deep-sea or turbidly water application, additional illumination is required. Figures 1 and 2 show the light absorption process in water.

The underwater imaging process is that, underwater optical cameras are usually equipped in watertight enclosures including a depth rated lens. Before reaching the scene of the underwater optical camera, the refraction causes the light rays coming

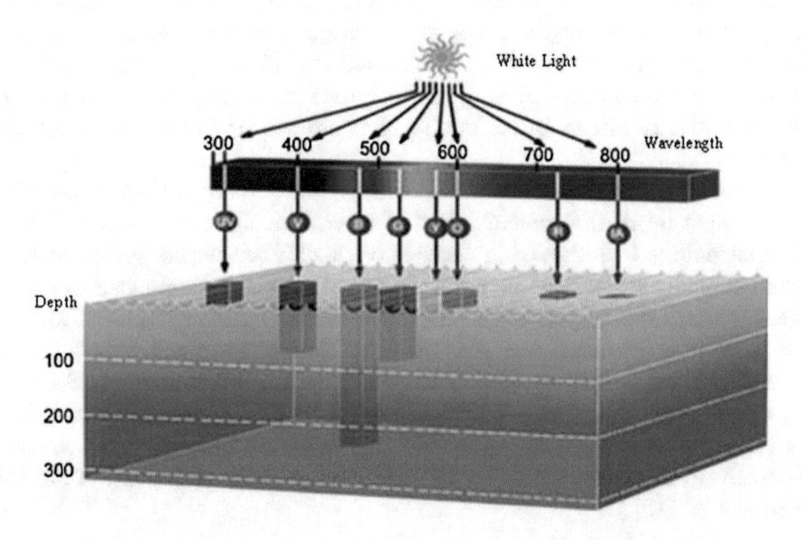

Fig. 1 The diagram shows the depth that light will penetrate in clear ocean water

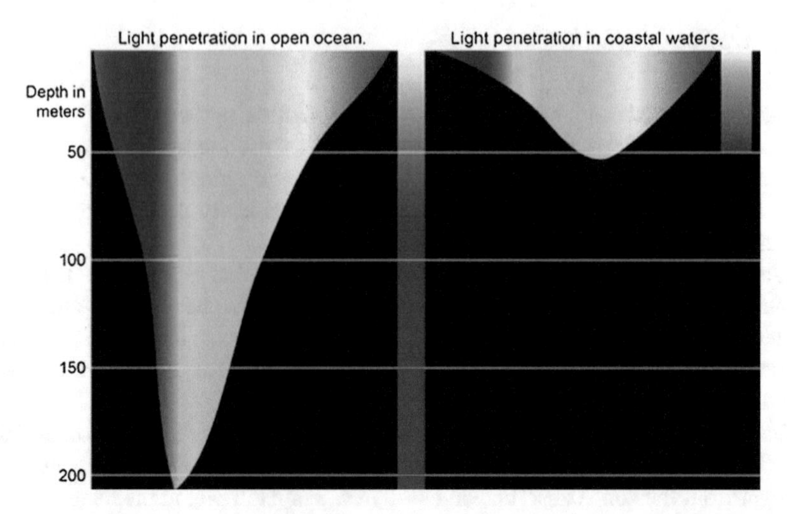

Fig. 2 NOAA basic illustration of the depth at which different color of light penetrates ocean waters. Water absorbs warm colors like *reds* and *oranges* (known as long wavelength light) and scatters the cooler colors (known as short wavelength light)

the scene to bent as they pass from water to glass and then from glass to air. The refraction modifies the apparent size and position of objects [16, 18].

When the light (or a photon) hits a particle suspended in water, its original path is deflected by the medium. Depending on the angle of impurities, the light ray is deviated, this phenomenon is known as forward scattering or backscattering. Forward scatter always occurs when the angle of deflection is small, resulting in image blurring and contrast reduction. Backscatter occurs when the light from the light source is reflected to the camera before reaching the object to be illuminated. Backscatter may cause bright points in the image usually known as marine snow [16]. The main issue of backscatter, also named as veiling light, is that it can highly reduce the image contrast, causing serious problems in underwater imaging systems. The referred effects of backscatter, forward scatter and refraction are illustrated in Fig. 3.

Both forward scatter and backward scatter are depending on the scope of illuminated water inside the camera's field of view. The absorption is caused by the electromagnetic waves traversing water to be quickly attenuated. Furthermore, the spectral components of light are absorbed quickly. Therefore, in clean water, long wavelength (red band light) is lost at first. In turbid water or in places with high concentration of plankton, red light may be better transmitted than blue light. Consequence, two problems are noticed, which are important problems for optical imaging and computer vision processing systems. Firstly, the usage of artificial light is needed in most cases and dramatically limits the distance at which objects are perceived. Secondly, the colours are distorted and the perception of the scene can be altered [16].

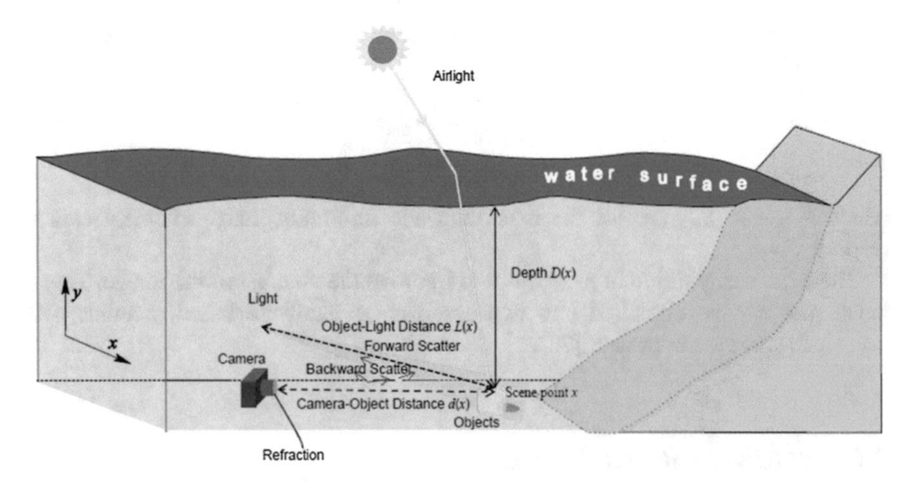

Fig. 3 Example of backward scatter, forward scatter and refraction

For underwater optical imaging system design, the better solution is to separate the illumination sources and the underwater optical camera, so that the backscattered light is separated from the observer as much as possible. In general, light source is separated from the camera as much as possible by other small underwater equipment (about 3–5 m). Another approach which reduces the effect of backscatter is to use gated viewing technology, which is used to emit a short pulse of light, and the camera is opened only when the light pulse passes the desired viewing distance. Therefore, the backscatter effect from the turbidity is not showed on the image. The third approach to increase visibility is to take polarized filters, cross polarized between the illumination and the underwater camera.

3 Challenges of Underwater Imaging Systems

As mentioned before, the main challenge working with the results of underwater imaging system from both rapid decay of signals of absorption, which leads to poor signal to noise feedbacks, and blurring caused by strong scattering by the water itself and constituents within, epically particulates. To properly address these issues, knowledge of in-water optical properties and their relationship to the image formation can be exploited in order to restore the imagery to the best possible level [19].

The processing of improving a degraded image to visibly look better is called image enhancement or image quality improvement [20]. It is explained that, due to the effects of optical or acoustic backscatter, the images in a scattering medium have low contrast. By improving the image contrast, it is expected to increase the visibility and discern more details. There are different definitions of measuring image contrast. One of the common definitions for image contrast c is the Michelson formula [21]:

$$c = \frac{I_{\max} - I_{\min}}{I_{\max} + I_{\min}} \tag{1}$$

where I_{max} and I_{min} are for the maximum and minimum image intensity values respectively.

There are many different techniques to improve the contrast of the image. These techniques can be classified into two approaches: hardware based methods and non-hardware base approach [22].

3.1 *Hardware Based Approach*

Hardware based approach requires special equipment; two common examples include polarization and range-gated imaging.

(1) **Polarization**

Light has three properties, that is, intensity, wavelength, and polarization. The human vision system and some animals can detect polarization and use it in many different ways such as enhancing visibility [23]. Natural light is initially unpolarized. However, light reaching to a camera often has biased polarization due to scattering and refection. Light polarization coveys different information of the scene. Inspired by animal polarization vision, a polarization imaging technique has been developed. To collect light polarization data, polarization sensitive imaging and sensing systems are required [24].

Preliminary studies showed that the back-scatter light can be reduced by polarization. Some studies assume the reflected light from the object is significantly polarized rather than the back scatter and in some other studies the contrary is assumed. Also, in some studies active illumination, a polarized light source is used [25], whereas in other study passive illumination, ambient light is used for imaging. Polarization difference imaging (PDI) method process the intensity of two images obtained at two orthogonal polarizations. Schechner et al. introduced a method which is based on the physical model of visibility degradations to recover underwater images using raw images through different states of polarizing filter. In this method visibility can be restored significantly, but remains some noise due to pixels falling on distant objects. A technique is developed to reduce the noise [21]. This method is developed to capture images faster, and as a result may be able to estimate a rough 3D scene structure [26].

(2) **Range-Gated Imaging**

Range-gated or time-gated imaging is one of the hardware methods to improve the image quality and visibility in turbid conditions [27]. In range-gated underwater

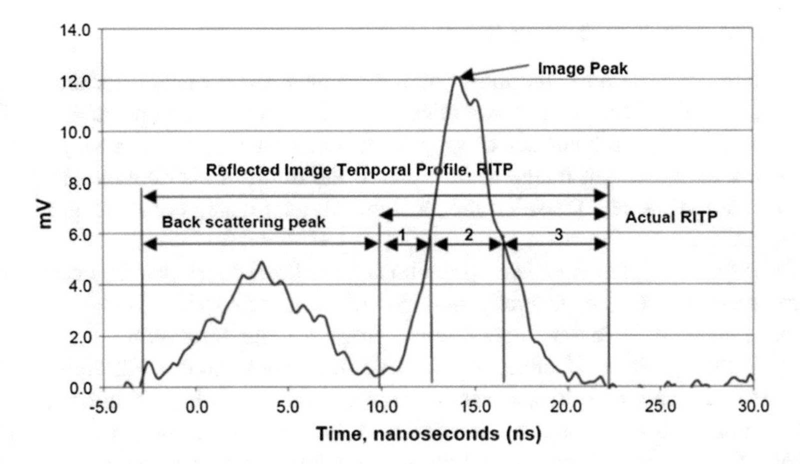

Fig. 4 The authorized copy of the timing plot of range-gated imaging system from [30]. Reflected Image Temporal Profile (RITP) in time domain, for clear water condition with attenuation coefficient, c = 0.26/m; absorption coefficient, a = 0:04/m 1. Front RITP, 2. Middle RITP, 3. Tail RITP

imaging system, the camera is adjacent to the light source, while the underwater target is behind the scattering medium [28]. The operation of range-gated system is to select the reflected light from the object that arrives at the camera and to block the optical back-scatter light [29].

Range-gated system includes a broad-beam pulse as the illumination source, a high speed gated camera and a synchronization gate duration control [29]. Tan et al. [28] presented a sample plot of the timing of range-gated imaging in their work. The authorized copy of the plot is shown in Fig. 4.

A range-gated process starts when the laser sends a pulse onto the object. As the light travels, the camera gate is closed. Thus, back-scattered light will not be captured. The fast electronic shutter of the gated camera is time delayed and only opens for a very short period of time. When the laser pulse returns to the camera after hitting the object, the camera gate opens. In this case, the camera is exposed only to the reflected light from the object. Once the laser pulse is over, the camera gate closes again. The opening or closing of the camera gate is based on the prior information about the object location [30].

3.2 *Non-hardware Based Approach*

In non-hardware based approach, no special imaging equipment is required and only digital image processing tools are utilized. Three common examples include histogram equalization, statistical modeling and unsharp masking.

(1) **Histogram Equalization (HE)**

Histogram equalization is the most common enhancement method for underwater image processing because of its simplicity and effectiveness. To operation of HE is to redistribute the probabilities of gray levels occurrences in such a way that the histogram of the output image to be close to the uniform distribution. Histogram equalization does not consider the content of an image, only the gray level distribution.

Different histogram equalization methods have been developed in recent years. These methods can be generally divided into two categories: global and local methods. Global histogram equalization processes the histogram of the whole image. Although it is effective, it has some limitations. Global HE stretches the contrast over the whole image, and sometimes this causes loss of information in dark regions. To overcome this limitation, a local HE technique was developed. Local HE uses a small widow that slides sequentially through every pixel of the image. Only blocks of the image that fall in this window are processed for HE and the gray level mapping is done for the center pixel of that window. Local HE is more powerful, but requires more computation. Local HE sometimes causes over enhancement is some parts of the image, and also increases the image noise. Some methods are developed to speed up the computation, such as partially overlapped histogram equalization and block based binomial filter histogram equalization.

(2) **Statistical Modelling**

Oakley and Bu [31] introduce a statistical based method, which using the standard deviation of the normalized brightness of an image to detect the presence of optical back scatter in a degraded image. It is assumed that the level of the optical back-scatter is constant throughout the image. This algorithm intends to find the minimum of a global cost function.

The proposed algorithm for optical backscatter estimation is to find the minimum value of a cost function that is a scaled version of the standard deviation of the normalized intensity. The key feature of this method is that it does not require any segmentation as it uses a global statistic rather than the sample standard deviation of small blocks.

The enhanced version of an image has the form:

$$\hat{I} = m(I - b) \tag{2}$$

where I is the degraded image, b is an estimate of the optical back-scatter contributed part of the image, is the modified image and m is the scaling parameter. The estimated value of optical back-scatter has been shown:

$$\arg\min\{S(b)\} \tag{3}$$

where

$$S(b) = \frac{1}{P}\sum_{p=1}^{P}\left(\frac{I_p - \bar{I}_p}{\bar{I}_p - b}\right)^2 GM\{(\bar{I}_p - b) : p = 1, 2, \ldots, P\} \tag{4}$$

where p is the pixel position, P is the total number of pixels, I is the degraded image, is the smooth version of the image, which is calculated by reclusive Gaussian filter.

(3) **Unsharp Masking (UM)**

Unsharp masking (UM) is the other common image enhancement method [32]. In this method the image is improved by emphasizing the high frequency components in the image [33].

The UM method is derived from an earlier photographic technique and involves subtracting the blurred version of an image from the image itself [33]. This is equivalent to adding a scaled high-pass filtered version of the image to itself [34] as shown in Eq. (5). The high pass filtering is usually done with a Laplacian operator [16].

$$y(m,n) = x(m,n) + \lambda z(m,n) \tag{5}$$

where $x(m, n)$ is the original image, λ is a constant, greater than 0, that changes the grade of sharpness as desired and $z(m, n)$ is the high-pass filtered version of the original image.

Although this method is easy to implement, it is very sensitive to noise and also causes digitizing effects and blocking artifacts. Different methods of UM have been introduced to mitigate these problems. Non-linear filters, such as polynomial and quadratic filters are used instead of the high pass filter.

4 Issues in Imaging Systems

One effect of the inherent optical properties (IOP) of ocean is that it becomes darker and darker with the water depth increases. As the water depth increases, the light from the sun is absorbed and scattered. For example, in the clean ocean water, the euphotic depth is less than 200 m [19]. In addition, the spectral composition of sunlight also changes with the water depth. Absorption is larger for long wavelengths (red color) than for short (green color); therefore, most of underwater images taken by natural light (sunlight) will appear blue or green on images or videos. Consequently, for the application of deep-sea or turbidly water, additional illumination is required.

4.1 Scattering

The volume scattering function describes the angular distribution of light scattered by a suspension of particles toward the direction at a wavelength. In the past years, many researchers in optical oceanography built instruments to measure the volume scattering function of sea waters [20, 22, 23, 35, 36]. Figure 5 shows three of Petzold's VSF curves displayed on a log-log plot to emphasize the forward scattering angles. Then instruments are a spectral response centered at $\lambda = 514\,\text{nm}$ with a bandwidth of 75 nm [37].

4.2 Absorption

When light penetrates the water, photons are either absorbed or scattered. While scattering redirects the angle of the photon path, absorption removes the photons from the light path. The absorption is highly spectrally dependent. In practice, it is hardly to measure the absorption rate [38–46].

Absorption by water is shown in Fig. 6. The blue wavelength is more highly absorbed than red wavelength.

The absorption rate of phytoplankton is shown in Fig. 7.

4.3 Color Distortion

Compare with common photographs, underwater optical images suffer from poor visibility owing to the medium. Large suspended particles cause scattering in turbid water. Color distortion occurs because different wavelengths are attenuated to different degrees in water. Meanwhile, absorption of light in water substantially

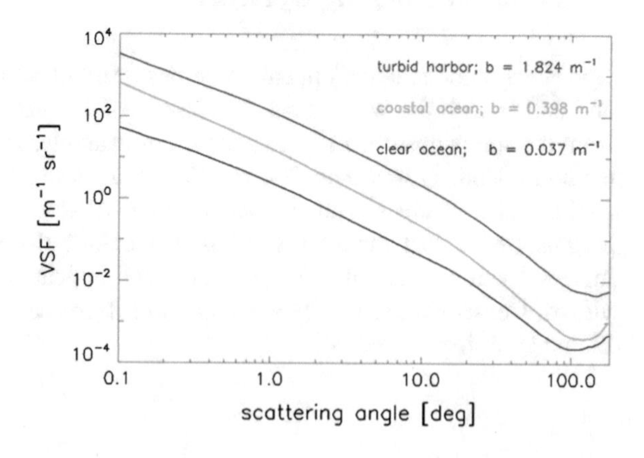

Fig. 5 Log-log plots of Petzold's measured volume scattering function from three different waters

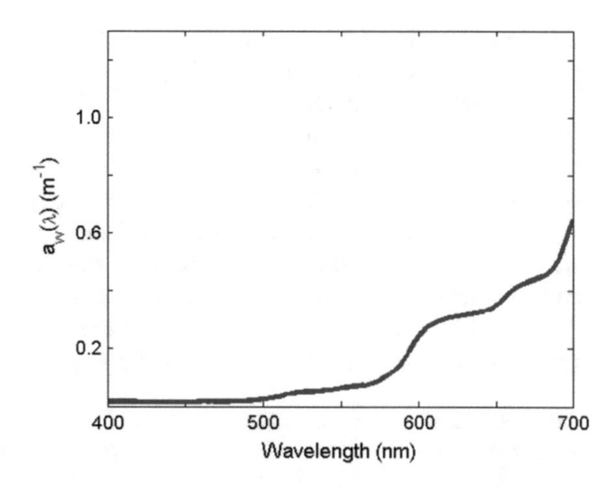

Fig. 6 Absorption spectrum for pure water

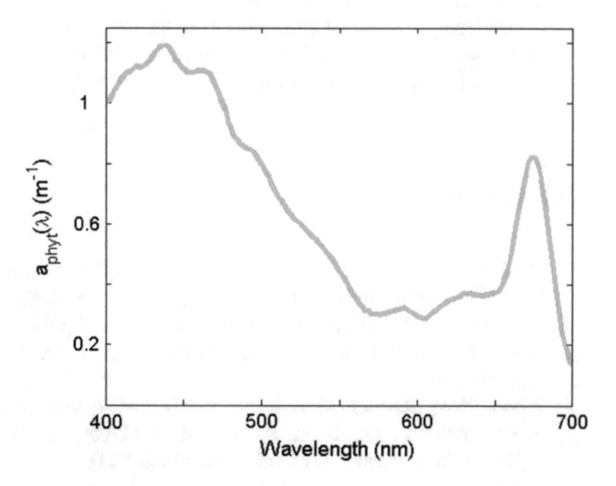

Fig. 7 Generic phytoplankton absorption spectrum for mixed algal composition

reduces its intensity. Furthermore, the random attenuation of light causes a hazy appearance as the light backscattered by water along the line of sight considerably degrades image contrast. So, underwater images contrast enhancement becomes more and more important [47, 48].

Other issues such as artificial lighting, camera reflection, blurring is also affecting the quality of underwater images [49–52].

5 Conclusions

There are some methods attempt to solve the underwater surveying vision problems by AUVs or Deep-sea Mining Systems (DMS). The field of underwater imaging is very new and much remains to be explored. Several long-term results would make a difference. Regarding future work, there are still several open problems which will require the development of new techniques.

On the one hand, to improve the existing underwater imaging system, thereby it can fully recover the real color of underwater images. On the other hand, for underwater optical system, we want to do some improvements and then apply them to some applications, such as underwater archeology, fish observation and so on.

Finally, we would like to combine the existing system with robots and make the robots to do some underwater exploration, such as underwater search, mineral exploration, and detection salvage and so on. It is an important future direction that improves monitoring underwater robots which not entirely dependent on intelligent machines but more dependent sensors and human intelligence.

Acknowledgments All of the authors have the same contribution to this paper. This work was supported by Grant in Aid for Foreigner Research Fellows of Japan Society for the Promotion of Science (No. 15F15077), Open Fund of the Key Laboratory of Marine Geology and Environment in Chinese Academy of Sciences (No. MGE2015KG02), Research Fund of State Key Laboratory of Marine Geology in Tongji University (MGK1407), Research Fund of State Key Laboratory of Ocean Engineering in Shanghai Jiaotong University (1315; 1510), and Grant in Aid for Research Fellows of Japan Society for the Promotion of Science (No. 13J10713).

References

1. Ocean observations, Wikipedia. http://en.wikipedia.org/wiki/Ocean_observations
2. Ocean Networks Canada. http://www.oceannetworks.ca/installations/observatories
3. O. NOAA Ocean Climate Observation Program. http://www.oco.noaa.gov/
4. Japanese ocean flux data sets with use of remote sensing observation. http://dtsv.scc.u-tokai.ac.jp/j-ofuro/
5. Woods Hole Oceanographic Institution. https://www.whoi.edu/
6. F. Muller-Karger, M. Roffer, N. Walker, M. Oliver, O. Schofield, Satellite remote sensing in support of an integrated ocean observing system. IEEE Geosci. Remote Sens. Mag. **1**(4), 8–18 (2013)
7. Data Buoy Cooperation Panel. http://www.jcommops.org/dbcp/
8. Underwater acoustic sensor networks. http://www.ece.gatech.edu/research/labs/bwn/UWASN/
9. J.P. Fish, *Sound Underwater Images: A Guide to the Generation and Interpretation of Side Scan Sonar Data*, 2nd edn. (Lower Cape Publishing, Oreleans, MA, 1991)
10. J.S.M. Rusby, J. Revie, Long-range mapping of the continental shelf. Mar. Geol. **19**(4), M41–M49 (1975)
11. P. Blondel, B.J. Murton, Handbook of Seafloor Sonar Imagery, Wiley-Praxis Series in remote sensing, ed. by D. Sloggett (Wiley, Chichester, 1997)
12. E. Belcher, W. Hanot, J. Burch, Dual-frequency identification sonar, in Proceedings of the 2002 International Symposium on Underwater Technology, pp. 187–192 (2002)
13. A. Davis, A. Lugsdin, High speed underwater inspection for port and harbor security using Coda Echoscope 3D sonar. Proc. MTS/IEEE OCEANS **3**(1-6), 2005 (2005)
14. BlueView Technologies. http://www.blueviewtech.com. Accessed 14 Dec 2013
15. R. Schettini, S. Corchs, Underwater image processing: state of the art of restoration and image enhancement methods. EURASIP J. Adv. Signal Process. **2010**, ID 746051, 1–14 (2010)

16. F. Bonin, A. Burguera, G. Oliver, Imaging system for advanced underwater vehicles. J. Marit. Res. **VIII**(1), 65–86 (2011)
17. Ocean Optics. www.oceanopticsbook.info
18. F. Bonin, A. Burguera, G. Oliver, Imaging systems for advanced underwater vehicles. J. Marit. Res. **VIII**(1), 65–86 (2011)
19. M. Chambah, D. Semani, Underwater color constancy: enhancement of automatic live fish recognition, in Proceedings of SPIE, pp. 157–168 (2004)
20. Ocean Optics. www.oceanopticsbook.info
21. T. Treibitz, Y.Y. Schechner, Instant 3descatter, in Proceedings of IEEE Conference on Computer Vision and Pattern Recognition, pp. 1861–1868, New York, U.S.A (2006)
22. H. Lu, Y. Li, L. Zhang, S. Serikawa, Contrast enhancement for images in turbid water. J. Opt. Soc. Am. **32**(5), 886–893 (2015)
23. D.M. Kocak, F.R. Dalgleish, F.M. Caimi, Y.Y. Schechner, A focus on recent developments and trends in underwater imaging. Technol. Soc. J. **42**(1), 52–67 (2008)
24. K.M. Yemelyanov, S.S. Lin, E.N. Pugh, N. Engheta, Adaptive algorithms for two-channel polarization sensing under various polarization statistics with nonuniform distributions. Appl. Opt. **45**(22), 5504–5520 (2006)
25. S.S. Lin, K.M. Yemelyanov, E.N.P. Engheta, Polarization enhanced visual surveillance techniques, in Proceedings of the 2004 IEEE International Conference on Networking Sensing and Control, pp. 216–221, Taiwan (2004)
26. Y.Y. Schechner, Y. Averbuch, Regularized image recovery in scattering media. IEEE Trans. Pattern Anal. Mach. Intell. **29**(9), 1655–1660 (2007)
27. T. Treibitz, Y.Y. Schechner, Active polarization descattering. IEEE Trans. Pattern Anal. Mach. Intell. **31**(3), 385–399 (2009)
28. C.S. Tan, A.L. Sluzek, T.Y. Jiang, Range gated imaging system for underwater robotic vehicle, in Proceedings of IEEE International Symposium on the Applications of Ferroelectrics, pp. 1–6 (2007)
29. C. Tan, G. Sluzek, D.M. He, A novel application of range-gated underwater laser imaging system (ulis) in near target turbid medium. Opt. Laser Eng. **43**, 995–1009 (2005)
30. H. Li, X. Wang, T. Bai, W. Jin, Y. Huang, K. Ding, Speckle noise suppression of range gated underwater imaging system. Appl. Opt. **38**(18), 3937–3944 (2009)
31. J. Oakley, H. Bu, Correction of simple contrast loss in color images. IEEE Trans. Image Process. **16**(2), 511–522 (2007)
32. F. Gasparini, S. Corchs, R. Schettini, Low quality image enhancement using visual attention. Opt. Eng. **46**(4), 0405021–0405023 (2007)
33. R.C. Gonzalez, R.E. Woods, *Digital Image Processing*, 3rd edn. (U.S.A, Pearson Prentice Hall, 2008)
34. Y. Tao, H. Lin, H. Bao, F. Dong, G. Clapworthy, Feature enhancement by volumetric unsharp masking. Visual Comput. **25**(5–7), 581–588 (2009)
35. R. Schettini, S. Corchs, Underwater image processing: state of the art of restoration and image enhancement methods. EURASIP J. Adv. Signal Process. **746052** (2010)
36. R. Schettini, S. Corchs, Enhancing underwater image by dehazing and colorization. Int. Rev. Comput. Softw. **7**(7), 3470–3474 (2012)
37. C.O. Ancuti, T. Haber, P. Bekaert, Enhancing underwater images and videos by fusion, in Proceedings of IEEE Conference on Computer Vision and Pattern Recognition, pp. 81–88 (2012)
38. R. Fattal, Signal image dehazing, in SIGGRAPH, pp. 1–9 (2008)
39. K. He, J. Sun, X. Tang, Single image haze removal using dark channel prior. IEEE Trans. Pattern Anal. **33**(12), 2341–2353 (2011)
40. K. He, J. Sun, X. Tang, Guided image filtering, in Proceedings of the 11th European Conference on Computer Vision, vol. 1, pp. 1–14 (2010)
41. B. Ouyang, F.R. Dalgleish, F.M. Caimi, Image enhancement for underwater pulsed laser line scan imaging system. Proc. SPIE **8372**, 83720R (2012)

42. W. Hou, D.J. Gray, A.D. Weidemann, G.R. Fournier, Automated underwater image restoration and retrieval of related optical properties, in Proceedings of IEEE International Symposium of Geoscience and Remote Sensing, pp. 1889–1892 (2007)
43. S. Serikawa, H.M. Lu, Underwater image dehazing using joint trilateral filter. Comput. Electr. Eng. **40**(1), 41–50 (2014)
44. M. Chambah, D. Semani, A. Renouf, P. Courtellemont, Underwater color constancy: enhancement of automatic live fish recognition. Proc. SPIE **5293**, 157–168 (2004)
45. A. Rizzi, C. Gatta, D. Marini, A new algorithm for unsupervised global and local color correction. Pattern Recogn. Lett. **24**, 1663–1677 (2003)
46. J.Y. Chiang, Y.C. Chen, Underwater image enhancement by wavelength compensation and dehazing. IEEE Trans. Image Process. **21**(4), 1756–1769 (2012)
47. H. Lu, Y. Li, Underwater image enhancement method using weighted guided trigonometric filtering and artificial light correction. J. Visual Commun. Image Represent. **38**, 504–516 (2016)
48. Artificial light correction. Mar. Sci. Bull. **5**(1), 16–23 (2003)
49. D.M. Kocak, F.R. Dalgleish, F.M. Caimi, Y.Y. Schechner, A focus on recent developments and trends in underwater imaging. Mar. Technol. Soc. J. **42**(1), 52–67 (2008)
50. R.W. Preisendorfer, *Hydrologic Optics* (NOAA, PMEL, 1976)
51. E. Peli, Contrast in complex images. J. Opt. Soc. Am. **7**(10), 2032–2040 (1990)
52. M. Petrou, P. Bosdogianni, *Image Processing* (Wiley Press, The fundamentals, 1999)

Fault Diagnosis and Classification of Mine Motor Based on RS and SVM

Xianmin Ma, Xing Zhang and Zhanshe Yang

Abstract A fault diagnosis method that based on Rough Sets (RS) and Support Vector Machine (SVM) is proposed, because of the diversity and redundancy of fault data for the mine hoist motor. RS theory is used to analyze the stator current fault data of mine hoist machine in order to exclude uncertain, duplicate information. For getting the optimal decision table, the equivalence relationship of positive domains of between decision attributes and different condition attributes is analyzed in the decision tables to simplify condition attributes. The optimal decision table is as the SVM input samples to establish the SVM training model. And the mapping model which reflects the relation of the characteristics between condition attribute and decision attribute is obtained by SVM training model in order to realize the fault diagnosis of the mine hoist machine. The simulation results show that the fault diagnosis method based on RS and SVM ca accuracy of fault diagnosis.

Keywords Fault diagnosis · Fault classification · Mine hoist motor · RS · SVM

1 Introduction

Mine hoist is one of the "big four operating equipment" of coal mine and is called "throat" of mine, because it is the only hub connected to ground and underground. Mine hoist motor often runs under condition of frequent starting up, rotation, braking and variable loads, so motor often occurs to fault. If mine hoist motor can not be guaranteed to operate normally, not only affects the operation of the entire system, and also threaten the safety of personnel [1].

X. Ma · X. Zhang (✉) · Z. Yang
College of Electrical and Control Engineering, Xi'an University of Science & Technology, Shaanxi, China
e-mail: 1515228151@qq.com

X. Ma
e-mail: maxianminphd@163.com

H. Lu and Y. Li (eds.), *Artificial Intelligence and Computer Vision*,
Studies in Computational Intelligence 672, DOI 10.1007/978-3-319-46245-5_2

Common types of motor fault have rotor fault, stator fault, bearing fault. The practice results show that the rotor fault rate is the highest, and the rotor bar breaking fault is the most common in rotor fault, and the rotor bar breaking fault reaches more 48.4 % [2]. When the rotor of mine hoist machine goes wrong, the stator current will produce fault frequency characteristic component corresponding with the rotor fault [3]. The Fast Fourier Transform is used to analysis the spectrum of stator current in order to obtain mine hoist machine rotor fault data. The common set theories have classical set theory, fuzzy set theory and RS theory. RS becomes the most commonly used set theory, because membership function of RS can be obtained directly in the processing data without any additional information, so it has more stronger objective analysis capabilities and fault tolerance. Equivalence relation of domain of RS theory is used to judge the collection which is made up by the similar elements in positive domain of concept, these collections are used to establish decision table of mine hoist machine fault data, and the decision table is used to search rules to predict and classify the new data. But these rules are mostly dependent on the logical reasoning of knowledge base, and there is no any relationship can be found among these rules, so the diagnosis rate is low and the efficiency is not high for this method [4, 5]. Therefore, in this paper, the intelligent algorithms are introduced, because of the limitations of expert system and the existence of empirical risk minimization principle of artificial neural network, the SVM algorithm with the feature of achieving structural risk minimization principle is used and it can also solve the small sample problem. Then the optimal decision table after reduction is as the input samples of SVM to diagnose and classify the mine hoist machine fault [6].

2 Stator Current Fault Analysis

When the rotor of mine hoist machine goes wrong, the stator current will produce fault frequency characteristic component corresponding with the rotor fault, the side frequency characteristics of the fault can be shown [7]:

$$f_b = (1 \pm 2ks)f_1 \tag{1}$$

where s is slip; f_1 is power supply frequency (50 Hz).

When the rotor of mine hoist machine goes wrong, the expression of stator current can be written:

$$\begin{aligned} i = I\cos(wt - \varphi_0) + I_1\cos[(1 - 2s)wt - \varphi_1] \\ + I_2[(1 + 2s)wt - \varphi_2] \end{aligned} \tag{2}$$

where I_1, I_2 are the amplitude of the stator current fault component feature after the rotor broken bars; φ_1, φ_2 are the characteristic initial phase corresponding to fault current component.

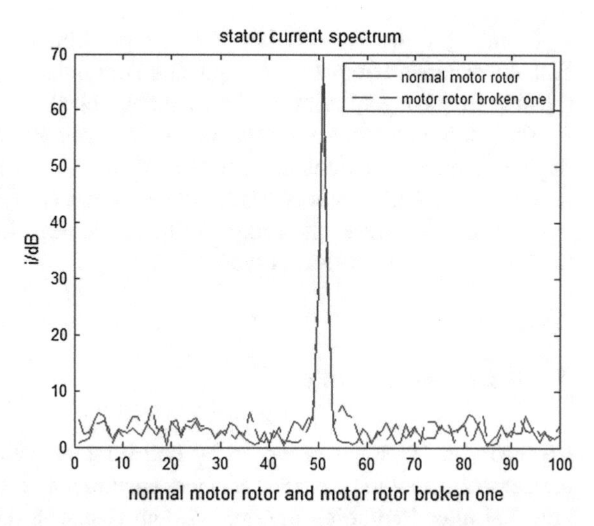

Fig. 1 Spectrum of normal rotor and rotor broken one

The simulation spectrum of stator current is obtained. But the stator current amplitude $(1 \pm 2ks)f_1$ is small, in order to prevent the component is affected by the magnitude of the fundamental current, the Hamming window function is used. When the slip is 0.02, the simulation spectrum of normal rotor and rotor broken one is shown in Fig. 1. The solid line represents stator current spectrum of motor rotor broken bars, the dotted line represents stator current spectrum of normal motor rotor.

From the Fig. 1 can be seen that when the rotor broken one the side frequency (the fault characteristic frequency components) of fundamental appears on the alongside. The side frequency appears on the frequency of 48 and 52 Hz.

When the motor rotor broken two, the stator current spectrum is shown in Fig. 2.

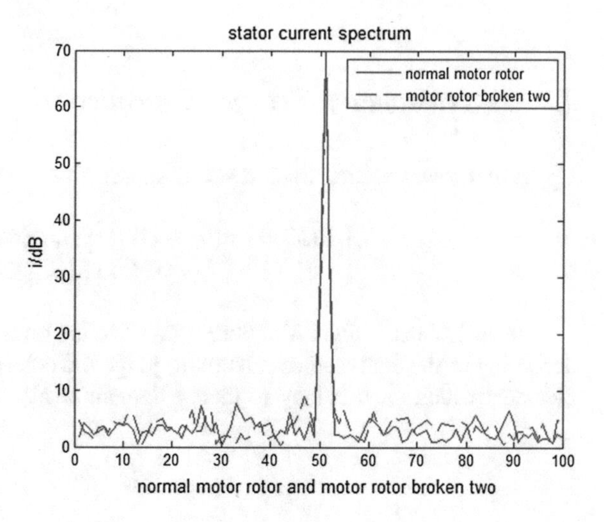

Fig. 2 Spectrum of normal rotor and rotor broken two

From Fig. 2 can be seen that when the rotor broken two the side frequency (the fault characteristic frequency components) of fundamental appears on the alongside, the side frequency appears on the frequency of 48, 52, 47, and 53 Hz.

The fault data that is extracted in spectrogram maybe has the features of duplicate, defects, redundancy, it will affect the result of diagnosis. In order to remove redundant condition attributes, eliminate duplicate samples to obtain the optimal decision table, the rough set theory is used to have the data sample pre-treatment before fault data is diagnosed.

3 RS Basic Theory

RS theory in 1982 was proposed by Polish mathematician Pawlak, RS theory is a mathematical tool to be used to analyze uncertain and redundant data, then to reveal potential rules by finding the implicit knowledge [8–11].

3.1 Knowledge Base and no Clear Relationship

Setting a non-empty finite set (theory domain)U, for any subset $X \subseteq U$ is known as the a concept or category of theory of domain, R is gens equivalence relation of $U, K = (U, R)$ is a knowledge base or approximate space, for an equivalence relation P, if $P \subseteq R$ and $P \neq \phi$, then all the intersection $(\cap P)$ of equivalence relation is also the equivalence relation on the theory of domain U, and the intersection is not clear on the relationship for an equivalence relation P, denoted $IND(P)$, further:

$$[X]_{IND(P)} = \bigcap\nolimits_{\forall R \in P} [X]_R \quad \forall x \in U \tag{3}$$

3.2 Approximation Set and Dependence

Upper and lower approximation set of subset $X \in U$ can be defined:

$$\begin{cases} \bar{R}(X) = \{x | (\forall x \in U) \wedge ([X]_R \cap X \neq \phi)\} \\ \underline{R}(X) = \{x | (\forall x \in U) \cap ([X]_R \subseteq X)\} \end{cases} \tag{4}$$

The positive domain of X about R s equal to the lower approximation set about$R$, according to the equivalence relation to judge the collection that is made up by the elements which must belong to theory domain of X.

Given $IND(K) = \{IND(P)|\phi \neq P \subseteq S\}$, it represents all equivalence relation in the knowledge base $K = (U, R)$, $\forall P, Q \in IND(K)$, the both exist the dependence of knowledge that is denoted:

$$\gamma_P(Q) = \frac{|pos_P(Q)|}{|U|} \tag{5}$$

where $pos_P(Q)$ is positive domain of Q about P.

3.3 Attribute Reduction

RS reduction is divided into attribute reduction and attribute value reduction, the attribute reduction is more complicated, methods of attribute reduction have resolution matrix reduction method and data analysis reduction method, etc.

Resolution matrix reduction method, in information systems $S = (U, B, V, f)$ of decision collection, among $B = C \cup D$ is collection of attributes, C is condition property, D is decision properties, V is value of the property, f expressed a kind of mapping: $U \times B \to V$, commonly used distinguish matrix to be expressed for:

$$M_D(i,j) = \begin{cases} \{b_k \in B \wedge b_k(x_i) \neq b_k(x_i)\} & d(x_i) \neq d(x_j) \\ 0 & d(x_i) = d(x_j) \end{cases} \tag{6}$$

where i expresses line, j says column, $i, j = 1, 2, 3, \ldots, n$, $M_D(i,j)$ represents elements of resolution matrix.

The resolution function is only defined by M_D, attributes $b \in B$, if $b(x, y) = \{b_1, b_2 \ldots, b_k\} \neq \phi$, specified a resolution function $b_1 \wedge b_2 \wedge \ldots \wedge b_k$, using $\sum b(x, y)$ to express it.

Data analysis reduction method, according to the information of the decision table (U, B) to carry on attribute reduction of attribute set B in turn, when a property is reduced to check the decision table whether to generate new rules, if it not to generate new rules, then the property can be reduced, or can not be reduced.

Delimited $r \in R$, if $IND(R) = IND(R - \{r\})$, then r is irreducible knowledge for R, if $P = (R - \{r\})$ is independent, P is a reduction about R, all irreducible relationship is called nuclear in R, denoted $CORE(R)$.

P and R are all equivalent relation cluster:

$$POS_{IND(P)}(IND(Q)) = pos_{IND(P - \{R\})}(IND(Q)) \tag{7}$$

If $R \in P$, then Q can be reduced for P, otherwise Q can not be reduced about P, equivalence relation set of all Q not about to go in P is called nuclear about Q for P, denoted $CORE_Q(P)$.

3.4 Data Decentralization

RS theory is applied only to deal with discrete data, but the collected data in the actual project is mostly continuous data, continuous attribute needs to be discrete into limited semantic symbol before to realize the processing of RS of continuous data attributes. The discrete methods commonly have discrete equidistant method, equal frequency method, and minimum entropy method, etc. But the easiest discrete way is dependent on the user own experience and knowledge to divided the area of continuous attributes into a plurality of are not mutually superimposed interval. Although rough set theory can be used to classify the new data according to the potential rules of the data reduction, this rule is more dependent on the logical reasoning of knowledge base and there will be no rules can be found among these rules, but speed is low and efficiency is not high for this diagnosis method. And when the rough set algorithm is introduced, the large number of sample data is reduced, because of the data sample is too small, the false positive rate will be greatly increased for expert systems with limitations and the artificial neural network with the principle of minimum empirical risk. Therefore, the Support Vector Machine (SVM) algorithm is applied with the principle of structural risk minimization, this algorithm can better be used to diagnose and classify small samples, nonlinear data.

4 Classification Principle of SVM Theory

SVM gets minimal actual risk and constructs statistical learning machine of optimal hyperplane based on structural risk minimization principle, SVM topology is determined by SV, and it can solve the issues which not easy to distinguish, such as small sample, non-linearity and low-dimensional space [11–15].

SVM is proposed from the case of linearly separable of the optimal hyperplane. There are training samples are assumed $E=\{(x_i,x_j),i=1,2,\ldots,n\},x\in R^d$, $y_i\in\{1,2,\ldots,k\}$, they can be correctly classified categories by established hyperplane, and the sample set should satisfy:

$$y_i[(w\cdot x_i)+b]-1\geq 0, i=1,2,\ldots,n \tag{8}$$

where w is weight, b is threshold.

The classification interval distance in this case from the above formula is $2/\|\omega\|$, when $\|w\|^2$ is minimized to get the maximum hyperplane. The Lagrange multipliers are used to solve objective function that establishes optimal hyperplane under $\sum_{i=1}^{n}\alpha_i y_i=0$ and $\alpha_i\geq 0$ (α_i is Lagrange multipliers $i=1,2,\ldots,n$).

$$\sum_{i=1}^{n}\alpha_i-\frac{1}{2}\sum_{i=1}^{n}\sum_{i=1}^{n}\alpha_i\alpha_j y_i y_j(x_i,x_j) \tag{9}$$

When α_i gets optimal solution, then optimal decision of classification function can be gotten:

$$F(x) = \operatorname{sgn}\left\{ \sum_{i=1}^{n} \alpha_i^* y_i (x_i \cdot x) + b^* \right\} \tag{10}$$

where $\operatorname{sgn}\{\cdot\}$ is sign function.

For linear inseparable issues, a slack variable $\xi_i \geq 0$ on the basis of linear problem is introduced, sample collection should meet:

$$y_i[(w \cdot x_i) + b] - 1 + \xi_i \geq 0, i = 1, 2, \ldots, n \tag{11}$$

When the formula satisfies the constraints $\sum_{i=1}^{n} \alpha_i y_i = 0$ and $0 \leq \alpha_i \leq C$, the α_i gets optimal solution, then optimal decision of classification function can be gotten:

$$F(x) = \operatorname{sgn}\left\{ \sum_{i=1}^{n} \alpha_i^* y_i (x_i \cdot x) + b^* \right\} \tag{12}$$

There are given a multivalued classifier that "one to one" combined with "One to many" algorithms, then the classification function of the multi-fault classifier can be established, the function can be written:

$$F^{\mathrm{m}}(x) = sign\left\{ \sum_{SV} \alpha_i^m y_i^m k(x_i, x) + b^m \right\} \tag{13}$$

5 Case Analysis

There are basic steps based on RS theory and SVM algorithm:

(1) The collected data is analyzed to do the normalization process
(2) The data after the normalization processing is discrete to form a decision Table
(3) For duplicate sample or redundant rules doing reduction in the decision Table
(4) Optimal decision table is gotten
(5) The optimal decision table is as input samples of SVM to establish SVM training model, samples have been treated by RS compare with the SVM simulation results of the samples which have not been treated by RS.

In this paper, these four fault types that rotor broken one (1), rotor broken two (2), rotor broken three (3) and rotor broken four (4) of mine hoist machine rotor are as an example for fault diagnosis, according to the spectrum analysis of stator current to extract the section frequency band data is as the mine hoist machine fault

Table 1 Mine hoist machine data

Sam.	a_1	a_2	a_3	a_4	a_5	a_6	a_7	a_8	Type
c1	0.045	0.032	0.018	0.764	0.104	0.029	0.079	0.011	1
c2	0.038	0.019	0.027	0.875	0.076	0.034	0.049	0.009	1
c3	0.052	0.029	0.048	0.796	0.087	0.073	0.073	0.006	1
c4	0.062	0.041	0.452	0.258	0.074	0.055	0.058	0.083	2
c5	0.023	0.061	0.163	0.843	0.239	0.042	0.042	0.007	2
c6	0.039	0.021	0.072	0.782	0.068	0.065	0.060	0.018	2
c7	0.044	0.072	0.033	0.812	0.169	0.028	0.022	0.017	2
c8	0.078	0.051	0.003	0.852	0.091	0.048	0.014	0.004	3
c9	0.057	0.042	0.053	0.524	0.259	0.100	0.097	0.031	3
c10	0.058	0.069	0.017	0.777	0.075	0.069	0.016	0.088	3
c11	0.022	0.036	0.054	0.432	0.226	0.255	0.551	0.020	4
c12	0.014	0.030	0.249	0.344	0.265	0.132	0.225	0.026	4
c13	0.037	0.026	0.020	0.701	0.053	0.044	0.213	0.005	4
c14	0.052	0.035	0.043	0.810	0.066	0.032	0.149	0.016	4

Table 2 Fault decision table

Sam.	a_1	a_2	a_3	a_4	a_5	a_6	a_7	a_8	D
c1	0	0	0	1	1	0	0	0	1
c2	0	0	0	1	0	0	0	0	1
c3	0	0	0	1	0	0	0	0	1
c4	0	0	1	1	0	0	0	0	2
c5	0	0	1	1	1	0	0	0	2
c6	0	0	0	1	0	0	0	0	2
c7	0	0	0	1	1	0	0	0	2
c8	0	0	0	1	0	0	0	0	3
c9	0	0	0	1	1	1	0	0	3
c10	0	0	0	1	0	0	0	0	3
c11	0	0	0	1	1	1	1	0	4
c12	0	0	1	1	1	1	1	0	4
c13	0	0	0	1	0	0	1	0	4
c14	0	0	0	1	0	0	1	0	4

data, such as $\leq 0.125f_1(a_1)$,$(0.125 \sim 0.5)f_1(a_2)$,$(0.625 \sim 0.75)f_1(a_3)$,$(0.875 \sim 1) f_1(a_4)$,$(1.125 \sim 1.5)f_1(a_5)$,$(1.625 \sim 1.75)f_1(a_6)$, $(1.875 \sim 2)f_1(a_7)$, $> 2f_1(a_8)$, the f_1 is speed frequency. Data is shown in Table 1, and the data has been normalized.

The data in Table 1 is dispersed by utilization frequency method, if the mine hoist machine occurs fault in the corresponding bands, the data is marked 1, otherwise the data is marked 0. Fault type is denoted *D*. The decision table is shown in Table 2.

Table 3 The first reduction of the decision table

Sam.	a_3	a_5	a_6	a_7	D
c1	0	1	0	0	1
c2	0	0	0	0	1
c4	1	0	0	0	2
c5	1	1	0	0	2
c6	0	0	0	0	2
c8	0	0	0	0	3
c9	0	1	1	0	3
c11	0	1	1	1	4
c12	1	1	1	1	4
c13	0	0	0	1	4

In the decision Table 2, the same fault type has duplicate samples c2, c3, the sample c3 is removed; duplicate samples c5, c7, the sample c7 is removed; duplicate samples c8, c10, the sample c10 is removed; duplicate samples c13, c14, the sample c14 is removed.

In the condition properties, a_1, a_2, a_3, a_8 are belong to the same state for all decision attribute, and they are unable to correctly distinguish the decision attributes, so these conditions attributes are eliminated. Decision table after reduction is shown in Table 3.

Theory domain is U = (c1, c2, c4, c5, c6, c8, c9, c11, c12, c13), condition property is (a_3, a_5, a_6, a_7) in the Table 3.

The equivalence of theory domain for condition attribute can be described:

$$U/C = \{\{c1\}, \{c2, , c6, c8\}, \{c5\}, \{c4\}, \{c9\}, \{c11\}, \{c12\}, \{c13\} \tag{14}$$

The equivalence of theory domain for decision property can be denoted:

$$U/D = \{\{c1, c2\}, \{c4, c5, c6\}, \{c8, c9\}, \{\{c11, c12, c13\}\} \tag{15}$$

The similar equivalence relationships can be wrote:

$$U/a_3 = \{\{c1, c2, c6, c8, c9, c11, c13\}, \{c4, c5, c12\}\} \tag{16}$$

$$U/a_5 = \{\{c2, c4, c6, c8, c13\}, \{c1, c5, c9, c11, c12\}\} \tag{17}$$

$$U/a_6 = \{\{c1, c2, c4, c5, c6, c8, c13\}, \{c9, c11, c12\}\} \tag{18}$$

$$U/a_7 = \{\{c1, c2, c4, c5, c6, c8, c9\}, \{c11, c12, c13\}\} \tag{19}$$

The positive domain of D about C is:

$$pos_C(D) = \{c1, c4, c5, c9, c11, c12, c13\} \tag{20}$$

$$\gamma_C(D) = \frac{|pos_C(D)|}{|U|} = \frac{7}{14} = 0.5 \tag{21}$$

The dependence of D for C is 0.5, so uncertain Sample is {c2, c6, c8}, they can be discarded.

The positive regions can be computed:

$$pos_{C-a_3}(D) = \{c4,\ c5,\ c12\} \neq pos_C(D) \tag{22}$$

$$pos_{C-a_5}(D) = \{c1,\ c5,\ c9,\ c11,\ c12\} \neq pos_C(D) \tag{23}$$

$$pos_{C-a_6}(D) = \{c9,\ c11,\ c12\} \neq pos_C(D) \tag{24}$$

$$pos_{C-a_7}(D) = \{c11,\ c12,\ c13\} \neq pos_C(D) \tag{25}$$

Therefore, reduction of the final condition property is (a_3, a_5, a_6, a_7).

The optimal decision table eventually can be gotten which is shown in Table 4.

The optimal decision table data samples {c1, c4, c5, c9, c11, c12, c13} are established matrix of the condition attributes and the fault types {1, 2, 3, 4} are established matrix of decision attribute, those matrix are as the input sample of SVM to train the SVM model.

Choosing different the inner product kernel functions to form different algorithms, there are four kernel functions are more commonly useful in the classification: linear kernel, polynomial kernel function, RBF kernel function and sigmoid kernel function. After several tests, the RBF is used.

$$k(x, y) = \exp\left[-\|x-y\|^2 / (2s)^2\right] \tag{26}$$

where x, y are training data; s is the width of the RBF. In the paper s takes 0.5, error penalty factor C takes 10.

Table 4 Optimal decision table

Sam.	a_3	a_5	a_6	a_7	D
c1	0	1	0	0	1
c4	1	0	0	0	2
c5	1	1	0	0	2
c9	0	1	1	0	3
c11	0	1	1	1	4
c12	1	1	1	1	4
c13	0	0	0	1	4

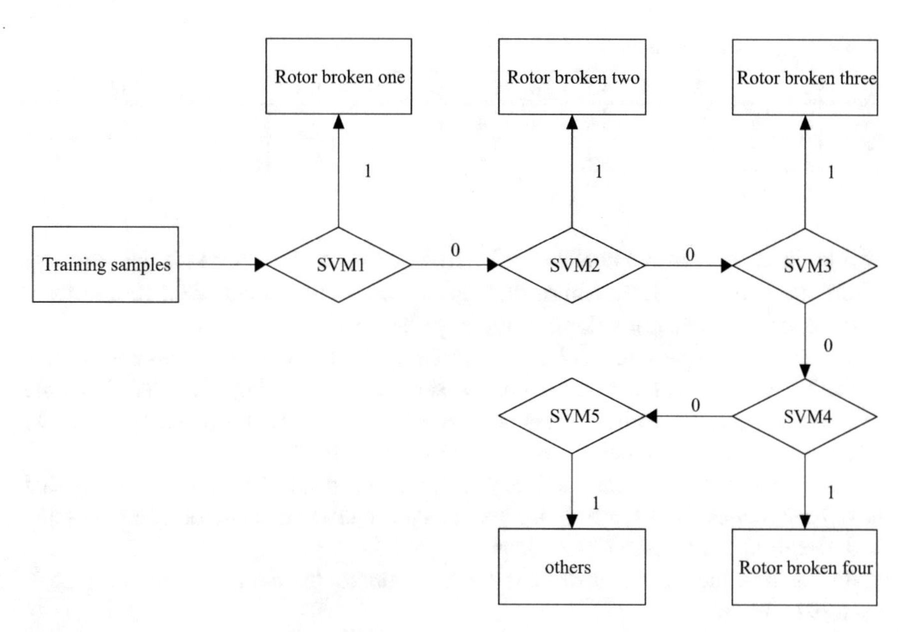

Fig. 3 The flowchart of multiple fault classifiers

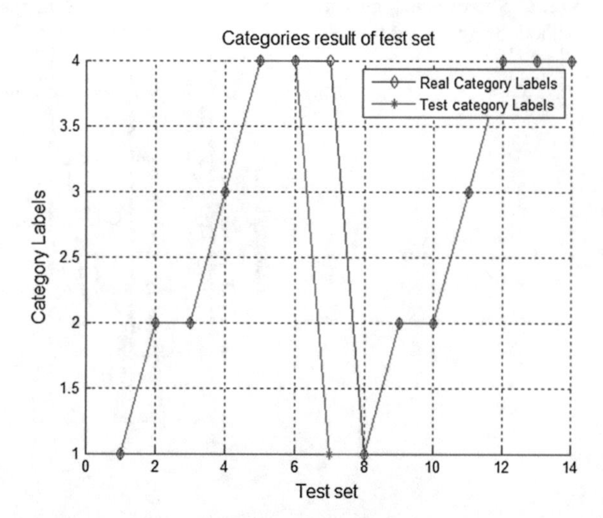

Fig. 4 SVM training results is treated by RS

The fault characteristic is as the input sample of multi-fault classifier, then the flowchart can be obtained of multiple fault classifier, as the Fig. 3 is showed.

If the output of discriminant $F^1(x)$ is 1, the sample belongs to class 1(rotor broken one), the training is finished; otherwise, training samples will go in the classifier 2 automatically, then the output of discriminant $F^2(x)$ is 1, the sample belongs to class 2 (rotor broken two), the test is finished, otherwise, test samples

Table 5 Analysis simulation result

L	1	2	3	4	5	6	7	8	9	10	11	12	13	14
M	1	2	2	3	4	4	**4**	1	2	2	3	4	4	4
N	1	2	2	3	4	4	**1**	1	2	2	3	4	4	4

will go in the classifier 3 automatically. And so on, for classifier k, if the output of discriminant $F^k(x)$ is 1, the sample belongs to class k, the output of the discriminant is 0,the sample does not belong to the any classifier.

Some test samples are selected, according to the above processing of RS reduction to carry on the reduction of test samples to get the optimal decision table of tested sample, the optimal decision table of tested sample is applied to the SVM training model, the simulation results is shown in Fig. 4.

The Table 5 is established to analyze simulation results of the test samples that have been treated by RS in Fig. 4. There L represents sample number; M represents real category; N represents test category.

In Table 5, there is an error classification sample, the classification accuracy is up to 92.857 %.

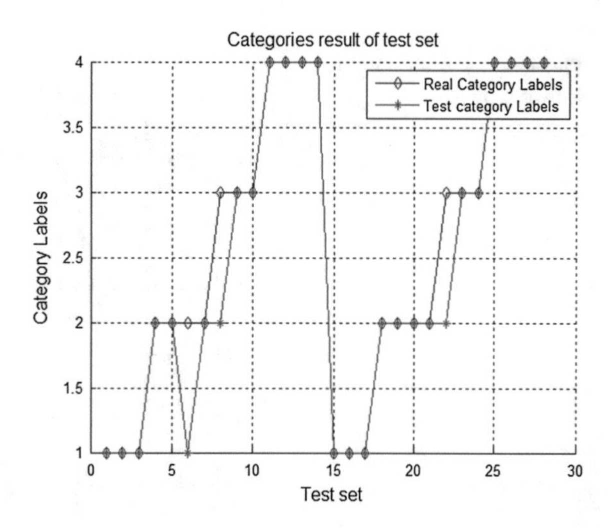

Fig. 5 SVM training results without being treated by RS

Table 6 Analysis simulation result

L	1	2	3	4	5	6	7	8	9	10	11	12	13	14
M	1	1	1	2	2	**2**	2	**3**	3	3	4	4	4	4
N	1	1	1	2	2	**1**	2	**2**	3	3	4	4	4	4
L	15	16	17	18	19	20	21	22	23	24	25	26	27	28
M	1	1	1	2	2	2	2	**3**	3	3	4	4	4	4
N	1	1	1	2	2	2	2	**2**	3	3	4	4	4	4

The SVM simulation result of test samples that had not been processed by RS is shown in Fig. 5.

The Table 6 is established to analyze simulation results of the test samples that have not been treated by RS in Fig. 5.

In Table 6, there are three error classification samples, the classification accuracy is up to 89.285 %.

The results of Table 5 compares to the results of Table 6, the accuracy of data which treated by RS to carry on fault diagnosis online has been greatly improved.

6 Conclusion

This article discusses the fault diagnosis method of mine hoist machine based on RS theory and SVM. There are some repeat and interference information of fault data which has been extracted from mine hoist machine. In order to reduce decision table and simplify diagnostic information, the RS theory is used to eliminate the uncertainty or repeated samples and to simply redundant attributes. The SVM is conducive to extract the data rapidly and raise the speed of fault diagnosis classification; in the simulation results, the fault diagnosis method based on RS and SVM is more accurate than the fault diagnosis method only based on SVM, so the fault diagnosis method has a certain practical value for the occasions of fault diagnosis needing online.

Acknowledgments National Natural Science Foundation of China (51277149), Project of Education Department of Shaanxi Province (14JK1472).

References

1. W. Dong, Study on monitoring and fault diagnosis system of mine hoist machine based on neural network. Shandong University of Science Thesis (2006)
2. Z. Yufang, Ma. Songlin, Development of virtual instrument system of motor fault diagnosis based on amperometric. Comput. Meas. Control **18**(3), 512–514 (2010)
3. L. Zhenxing, Y. Xianggen, Online diagnosis method of squirrel-cage induction motor rotor fault diagnosis based on Instantaneous power signal spectrum analysis[J]. Proc. CSEE **23**(10), 148–152 (2003)
4. Y. Li, H. Lu, L. Zhang, J. Zhu, S. Yang et al., An automatic image segmentation algorithm based on weighting fuzzy c-means clustering, in *Softer Computing in Information Communication Technology* (2012), pp. 27–32
5. M. Duoqian, L. Daoguo, *RS Theory, Algorithms and Applications* (Tsinghua University Press, Beijing, 2008)
6. H. Yuanbin, Fault diagnosis and classification of conveyor based RS theory. J. Chang'an Univ. **24**(1), 104–107 (2004)
7. L. Zhenxing, Y. Xianggen, Online diagnosis method of squirrel-cage induction motor rotor fault diagnosis based on Instantaneous power signal spectrum analysis. Proc. CSEE **23**(10), 148–152 (2003)

8. P. Wenji, L. Xingqi, Vibration Fault diagnosis of hydro-power units based on RSs and SVM [J]. Trans. China Electrotech. Soc. **21**(10), 118–122 (2006)
9. Y. Jun, W. Jianhua, Fault diagnosis system of induction motor based on SVM. Mechatron. Eng. **25**(1), 72–74 (2008)
10. W. Biao, D. Chanlun, W. Hao, et al., Research and Application of RSs and Fuzzy Sets (Electronic Industry Press, Beijing, 2008)
11. D. Ze, L. Peng, W. Xuehou, et al., Fault diagnosis of steam turbine based on RSs and SVM. J. North China Electric Power Uni. **23**(2), 79–82 (2008)
12. Y. Junrong, G. Xijin, Induction motor fault diagnosis based on least squares support vector machine. Comput. Meas. Control **21**(2), 336–339 (2013)
13. Z. Zhousuo, L. Lingjun, H. Zhengjia, Multi-fault classifier and application based on SVM. Mech. Sci. Technol. **23**(5), 0538–0601 (2004)
14. S.S. Keerthi, Efficient tuning of SVM hyper-parameters using radius/margin bound and iterative algorithms. IEEE Trans. Neural Netw. **3**(5), 1225–1229 (2002)
15. Z. Hongzhi, Yu. Xiaodong, Transformer fault diagnosis based on RSs and SVM. Transformer **45**(8), 61–65 (2008)

Particle Swarm Optimization Based Image Enhancement of Visual Cryptography Shares

M. Mary Shanthi Rani and G. Germine Mary

Abstract Due to the rapid growth of digital communication and multimedia applications, security becomes an important issue of communication and storage of images. Visual Cryptography is used to hide information in images; a special encryption technique where encrypted image can be decrypted by the human visual system. Due to pixel expansion the resolution of the decrypted image diminishes. The visual perception of a decrypted image can be enhanced by subjecting the VC shares to Particle Swarm Optimization based image enhancement technique. This improves the quality and sharpness of the image considerably. Suitable fitness function can be applied to optimize problems of large dimensions producing quality solutions rapidly. Results of the proposed technique are compared with other recent image enhancement techniques to prove its effectiveness qualitatively and quantitatively. The proposed algorithm guarantees highly safe, secure, quick and quality transmission of the secret image with no mathematical operation needed to reveal the secret.

Keywords Image enhancement · Particle swarm optimization · Image transmission · Visual cryptography · Information security · Secret sharing

1 Introduction

Information is the oxygen of the modern age. Valuing and protecting information are crucial tasks for the modern organization. In many applications information is sent in the form of images, as it requires less space and transmits more information. Due to the rapid growth of digital communication and multimedia applications,

M.M.S. Rani
Department of Computer Science and Applications,
Gandhigram Rural Institute—Deemed University, Dindigul, Tamil Nadu, India
e-mail: drmaryshanthi@gmail.com

G.G. Mary (✉)
Department of Computer Science, Fatima College, Madurai, Tamil Nadu, India
e-mail: germinemary@yahoo.co.in

H. Lu and Y. Li (eds.), *Artificial Intelligence and Computer Vision*,
Studies in Computational Intelligence 672, DOI 10.1007/978-3-319-46245-5_3

security turns out to be a significant issue of communication and storage of images. Cryptography is a tool for secure communication in the presence of adversaries. Visual Cryptography is used to hide information in images; a special encryption technique in such a way that encrypted image can be decrypted by the Human Visual System (HVS).

Visual Cryptography (VC) is "a new type of cryptographic scheme, which can decode concealed images without any cryptographic computation". Naor and Shamir first introduced VC technique in 1994 [1]. It is a powerful information security tool, which visually protects critical secrets from the view of hackers. In contrast to other security methods, which tend to conceal information by applying a mathematical transformation on secret; Visual Cryptography Scheme (VCS) stores the secret as an image. In this technique, a Secret Image (SI) is split up into n distinct meaningless images called shares; each of the shares looks like a group of random pixels and of course looks meaningless by itself [2]. Any single share does not reveal anything about the secret image. The secret image can be decrypted by stacking together all the n shares. The specialty of VC is that the secret can be retrieved by the end user by HVS without having to perform any complex computation.

In VC the reconstructed image after decryption process encounter a major problem. The image quality of the decrypted image is not exact as the original image due to pixel expansion, which is the greatest disadvantage of traditional VC. This pixel expansion results in the loss of resolution. The decrypted SI has a resolution lower than that of the original SI. This problem of contrast deterioration is overcome in this proposed method, by creating enhanced VC shares, by applying image enhancement technique using Particle Swarm Optimization (PSO) which improves the quality of the retrieved image.

Image enhancement techniques are used to highlight and sharpen image features such as to obtain a visually more pleasant, more detailed, or less noisy output image. The aim of image enhancement is to improve the interpretability or perception of information in the image for human viewers. There are many application-specific image enhancement techniques like Adaptive filtering, Median filter, Image Sharpening, Histogram Equalization, Contrast Enhancement etc. [3, 4]. Histogram equalization shows best result in most of the cases but if the image has wide light color, where adaptive histogram equalization may give better result [5].

Color image processing and enhancement is a more complicated process compared to black-and white images due to the presence of multiple color channels and the need to preserve the color information content while enhancing the contrast. Histogram equalization is popularly used and effectively proven method due to its simplicity and satisfactory performance [6].

Image enhancement of VC shares is different from other enhancement procedures as the image is made up of just two intensity values for each color channel. The evolutionary computation techniques like genetic algorithm and Particle Swarm Optimization (PSO) can be applied to enhance image contrast [7]. The aim of proposed method is to improve the quality of the reconstructed SI by minimizing the mean brightness error between the original and decrypted images. This is achieved by Image enhancement technique using PSO.

Furthermore, the proposed method is used to enhance the decrypted SI created from the VC shares of Floyd–Steinberg's Diffusion half toned image (FSDI) and is compared with VC created from the original image. The application of digital half toning techniques results in some reduction of the original image quality due to its inherently lossy nature and it is not possible to recover the original image from its halftone version [8].

The half toned image is converted to VC shares which amplify the contrast deterioration in the resultant decrypted SI. But it is exciting to note that the visual quality of the enhanced decrypted half toned SI is very good in comparison with the visual quality of the enhanced normal VC shares. By choosing the optimal intensities for RGB channels the quality of the retrieved SI is improved to a greater extent.

In this paper, we investigate a novel technique that aims at the reconstruction of SI with perfect quality, at the same time maintaining the confidentiality of the SI and preserving the characteristics of VC. The intensity transformation function uses local and global information of the input image and the objective function minimizes the difference in mean intensities of Original SI and VC shares.

This research paper is organized as follows. Section 2 describes the creation of VC shares from SI and FSDI color image. The proposed PSO based VC share enhancement method is explained in Sect. 3. Experiments results obtained by using color and gray images are described in Sect. 4, followed by discussion. A conclusion is drawn in Sect. 5.

2 Creation of FSDI and VC Shares

2.1 Creation of FSDI

Error diffusion techniques are used in most half toning transformations to change a multiple-level color image into a two level color image. The straightforward and striking concept of this technique is the diffusion of errors to neighboring pixels; thus, image intensity is not lost [9]. Error diffusion diagram is shown in Fig. 1, where f(m, n) represents the pixel at (m, n) position of the input image, g(m, n) is the quantized pixel output value and d(m, n) signify the sum of the input pixel values and the diffused errors. Error diffusion includes two main components. The first one is the thresholding block, in which the output g(m, n) is given by

$$\mathrm{g(m,n)} = \begin{cases} max, & if\ d(m,n) \geq t(m,n) \\ min, & otherwise \end{cases} \tag{1}$$

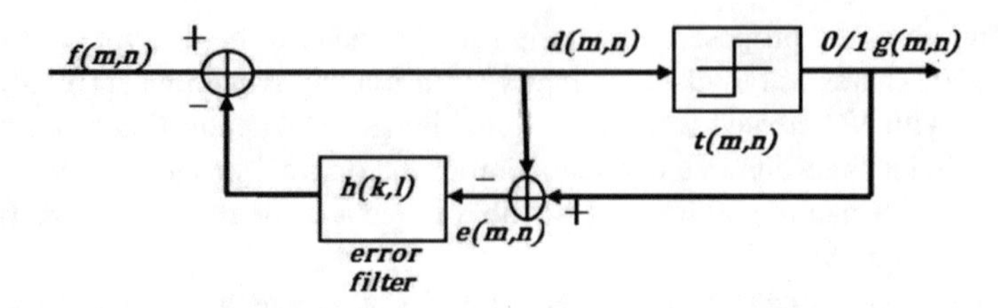

Fig. 1 Error diffusion block diagram

The threshold t(m, n) is position dependant. The second one is the error filter h (k, l). The input to the error filter e(m, n) is the difference between g(m, n) and d(m, n). Finally we calculate d(m, n) as

$$d(m, n) = f(m, n) - \sum_{k,l} h(k, l) e(m - k, n - l) \tag{2}$$

where *h(k, l)*ϵ *H* and *H* is a 2-D error filter.

The Floyd–Steinberg's algorithm adds the quantization error of a pixel onto its unprocessed adjoining pixels and processed. Floyd–Steinberg's distribution of pixel intensity is shown below.

$$H(k, l) = \begin{bmatrix} & * & 7/16\ldots \\ 3/16 & 5/16 & 1/16\ldots \end{bmatrix} \tag{3}$$

The algorithm scans the pixels in the image from left to right and from top to bottom, quantizing pixel values one by one. In the above matrix, the pixel being processed currently is indicated by star (*). Already processed pixels are represented by blank entries in the matrix. Every time the quantization error is shifted to the unprocessed neighboring pixels. Therefore if few pixels are rounded upwards, then the next few pixels have a high probability to be rounded downwards, thus maintaining quantization error close to zero [10].

The Secret color image is processed using Floyd–Steinberg's half toning method to produce the Floyd–Steinberg's Diffusion half toned image (FSDI) as follows

$$\mathrm{X}_{(\mathrm{m.n})} = \left[x^{R}_{(m,n)} \cdot x^{G}_{(m,n)} \cdot x^{B}_{(m,n)}\right] \in \{0, 255\} \tag{4}$$

where *m, n* lies between ((*1, h*), (*1, w*)) respectively.

Here X represents the pixel of the secret image and (m, n) represents the location of the pixel. The three binary bits $x^{R}_{(m,n)} \cdot x^{G}_{(m,n)} \cdot x^{B}_{(m,n)}$ represents the values for Red, Green and Blue color channels respectively. The histogram diagrams in Fig. 2 illustrate the frequency distribution of pixels in original SI and half toned SI.

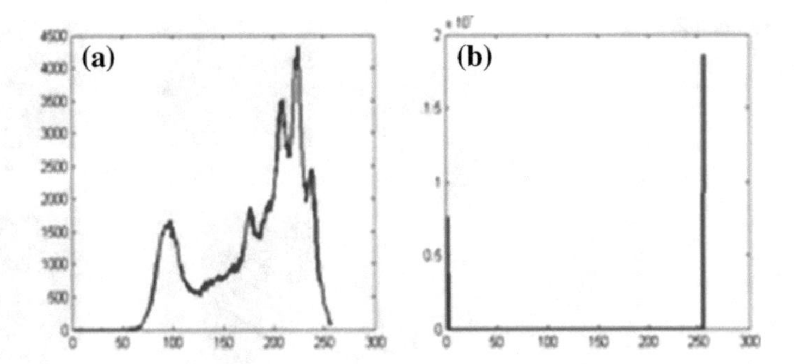

Fig. 2 Histogram comparison of Lenna image (**a**) Original SI (**b**) FSDI

2.2 Creation of VC Shares

Cryptography is a technique to scramble the secret message so that unauthorized users can't get a meaningful message. In conventional VCS, shares are formed as random patterns of the pixel. These shares look like meaningless noise. Noise-like shares do not stimulate the attention of hackers since it is complex to handle meaningless shares and all shares look alike [11]. In a (2, 2)—threshold color visual secret sharing scheme, let the SI be a half toned image ($X_{(m.n)}$) of size *mxn*. We use pixel value of '0' and '1' to represent black and color pixels (RGB) respectively. Naor and Shamir (1994) constructed the pixels of VC shares based on two basis matrices C_0 and C_1 as shown below.

$$C_0 = \{\text{all the matrices obtained by permuting the columns of} \begin{bmatrix} 1 & 1 & 0 & 0 \\ 1 & 1 & 0 & 0 \end{bmatrix}\} \tag{5}$$

$$C_1 = \{\text{all the matrices obtained by permuting the columns of} \begin{bmatrix} 1 & 1 & 0 & 0 \\ 0 & 0 & 1 & 1 \end{bmatrix}\} \tag{6}$$

where C_0 is used to represent shares of the black pixel and C_1 is used to represent shares of the color pixel.

As shown in Eq. (4), $X_{(m.\ n)}$ is split into 3 color channels (RGB). Two shares are created for every color channel depending on the intensity of pixel values of each color channel. Each pixel in every color channel is extended into a two 2 × 2 block to which a color is assigned according to the model presented in Fig. 3, and each block is composed of two black pixels and two color pixels. Figure 3 depicts the 2 × 2 blocks created for Red channel. The blocks are combined to form Share1 and Share2 for the red channel. In a similar way Share3 and Share4 for green channel and Share5 and Share6 for the blue channel are created [12].

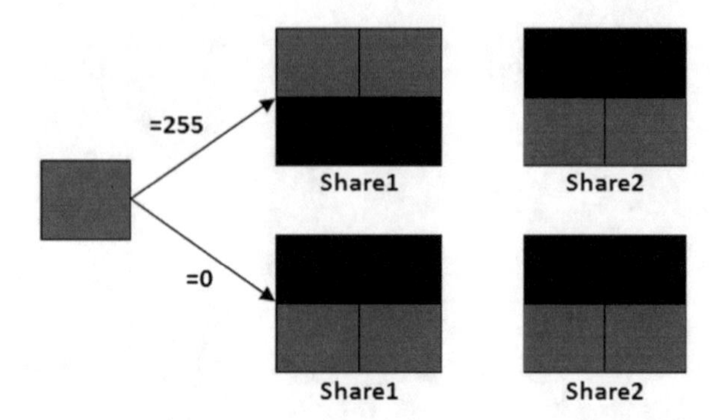

Fig. 3 Share creations for *red channel*

The six shares created will look like random dots and will not reveal any information since they have an equal number of black and color pixels. Finally, the shares of RGB, to be exact, the Shares 1, 3 and 5 are merged to form VC share1 and similarly Share2, Share4 and Share6 are merged to form VC share2 as in Fig. 4b, c.

In Fig. 4, the SI (a) is decomposed into two visual cryptography transparencies (b) and (c). When stacking the two transparencies, the reconstructed image (d) is obtained.

VC shares are created from original SI by following the same procedure with a small alteration. If a pixel intensity is >128 then the pixel is replaced by a 2 × 2 block of C_1 else by C_0. The contrast of the decrypted image is degraded by 50 % because of increase in the size of the image and human eyes can still recognize the content of the SI. SI is shown only when both shares are superimposed. Stacking shares represent OR operation in the human visual system. The size of the shares would increase considerably if quality and contrast are given priority [13].

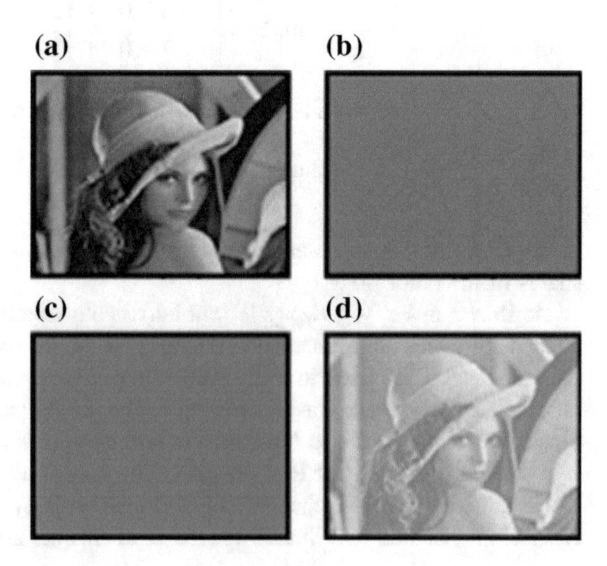

Fig. 4 **a** Color SI (FSDI), **b** Encrypted share1, **c** Encrypted share2, **d** Decrypted secret message

The important parameters of the scheme are:

m: the number of pixels in a share. This symbolizes the loss in resolution from the original image to the recovered one.
α: the relative difference in the weight between the combined shares that come from color and black pixel in the original image, i.e., the loss in contrast.
γ: the size of the collection of C_0 and C_1. C_0 refers to the sub-pixel patterns in the shares for a black pixel and C_1 refers to the sub-pixel patterns in the shares for a color pixel.

The Hamming weight $H(V)$ of the OR ed m-vector V is interpreted by the visual system as follows:

A color pixel is interpreted if $H(V) \leq d$ and black if $H(V) < d - \alpha.\ m$ for some fixed threshold $1 \leq d \leq m$ and a relative difference $\alpha > 0$.

3 Proposed Method

3.1 Basics of PSO

PSO is a simple, population-based, computationally efficient optimization method. It is a stochastic search based on a social-psychological model of social influence and social learning. In PSO individuals follow a very simple behavior of emulating the success of neighboring individuals. The aim of the PSO algorithm is to solve an unconstrained continuous minimization problem: find x^* such that $f(x^*) <= f(x)$ for all n-dimensional real vectors x. The objective function $f: Rn \rightarrow R$ is called the fitness function. PSO is swarm intelligence meta-heuristic inspired by the group behavior of animals, for example, bird flocks or fish schools. The population $P = \{p_1, \ldots, p_n\}$ of the feasible solutions is called a swarm and the feasible solutions $p_1, \ldots, p_n$ are called particles. The set R_n of feasible solutions is viewed as a "space" in PSO where the particles "move".

The particles in PSO fly around in a multidimensional search space and change their position with time. Every particle adjusts its position according to its own experience, and the knowledge of its neighboring particles, making use of the best position encountered by its neighbors and by itself. A PSO system merges local search and global search and attempts to explore regions of the search space and concentrate the search around a promising area to refine a candidate solution.

All the particles have a fitness value which is evaluated by the objective function to be optimized and a velocity which drive the optimization process and updates the position of the particles. First the group of particles is initialized randomly and it then searches for an optimal solution by updating through iterations. In all iterations, each particle is updated by following two "best" values. The best position reached by every particle thus far is the first value. This is known as *pbest* solution. The best position tracked by any particle among all generations of the swarm, known as gbest solution is the second value. These two best values are accountable to drive the particles to go to new better positions. After finding these two best values, a particle updates its velocity using (7) and position with the help of Eq. (8) [14].

$$v_i^{t+1} = W^t.v_i^t + c_1.r_1.\left(pbest_i^t - X_i^t\right) + c_2.r_2.\left(gbest^t - X_i^t\right) \tag{7}$$

$$X_i^{t+1} = X_i^t + v_i^{t+1} \tag{8}$$

where X_i^t and v_i^t denote the position and velocity of *i*th particle at time t, W^t is the inertia weight at *t*th instant of time, c_1 and c_2 are positive acceleration constants and r_1 and r_2 are random values in the range [0, 1], sampled from a uniform distribution. $Pbest_i$ is the best solution of ith individual particle over its flight path, gbest is the best particle attained over all generation so far.

In Eq. (7), the previous velocity v_i^t is an inertia component which remembers the direction of previous flight and prevents the particle from drastically changing its direction.

The cognitive component $c_1.r_1.\left(pbest_i^t - X_i^t\right)$ quantifies performances relative to past performances and remembers previous best position.

The Social component $c_2.r_2.\left(gbest^t - X_i^t\right)$ quantifies performances relative to neighbors (Fig. 5).

To begin the algorithm, the particle positions are randomly initialized, the velocities are set to 0, or to small random values, Swarm size (n_s), Particle dimension (n_x), Number of iterations (n_t), Inertia weight (w) and Acceleration coefficients (c1 and c2) are initialized [15].

In all generations, each particle is accelerated toward the particles previous best position and the global best position. For every particle, new velocity is calculated based on its current velocity, the distance from its previous best position, and the distance from the global best position. The next position of the particle in the search space is calculated based on new velocity. This process is then iterated n_t number of times or until a minimum error is achieved. The algorithm is terminated once the fitness values of the particles are achieved to the desired value or n_t iterations are over.

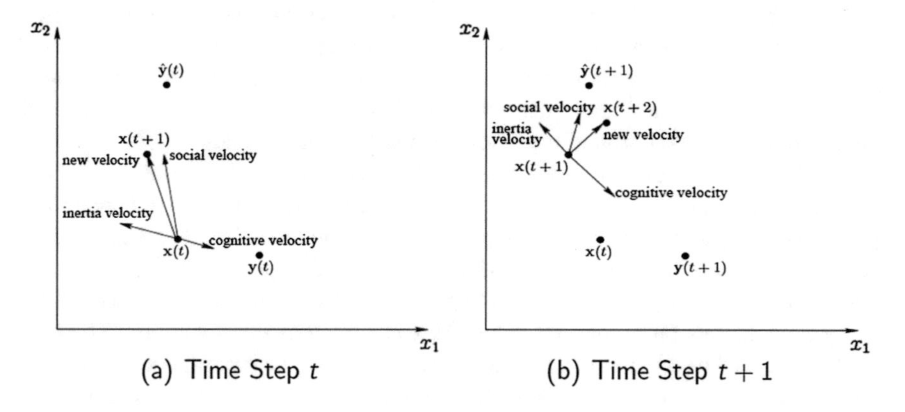

Fig. 5 Geometrical illustration of velocity and position updates for a single 2-D particle

3.2 Visual Cryptography Optimization Using PSO

The VC shares created from SI and FSDI SI (Fig. 4b, c) as discussed in Sect. 2, is sent across the network and the secret can be decrypted by the end user by overlapping the shares or by performing OR operation on the VC shares. The decrypted secret image (Fig. 6a, g) generally has low contrast and resolution because of pixel expansion. This can be resolved to a certain extent by means of image enhancement techniques. The disadvantage of the procedure is that these techniques have to be employed at the receiving end to the decrypted SI (merged VC shares). Moreover, the result of image enhancement of decrypted SI is not of superior-quality as expected and is shown in Fig. 6b–e, h–k. Hence in this proposed study, Image enhancement is done to VC shares using PSO, before it is sent across the network and at the same time secrecy of the shares are maintained.

The PSO algorithm for VC share enhancement is performed on Original VC shares as well as on FSDI VC shares by following the steps below.

Step 1: Read the secret color image (SI) and convert to FSDI using Floyd –Steinberg Dithering Algorithm using Eqs. (1) and (2).

Each pixel in FSDI can be represented as

$$\mathrm{X}_{(\mathrm{m.n})} = \left[x^{R}_{(m,n)}.x^{G}_{(m,n)}.x^{B}_{(m,n)}\right] \in \{\mathbf{0}, \mathbf{255}\}$$

Step 2: The FSDI/SI is decomposed into R, G, B channels. Every pixel in each of the RGB channel is expanded into a 2 × 2 block and two shares are created for each color channel as shown in Fig. 3.

$$[FSDI] - \text{split to}\, RGB - \left[FSDI_{r1}, FSDI_{g1}, FSDI_{b1}\right]\left[FSDI_{r2}, FSDI_{g2}, FSDI_{b2}\right]$$

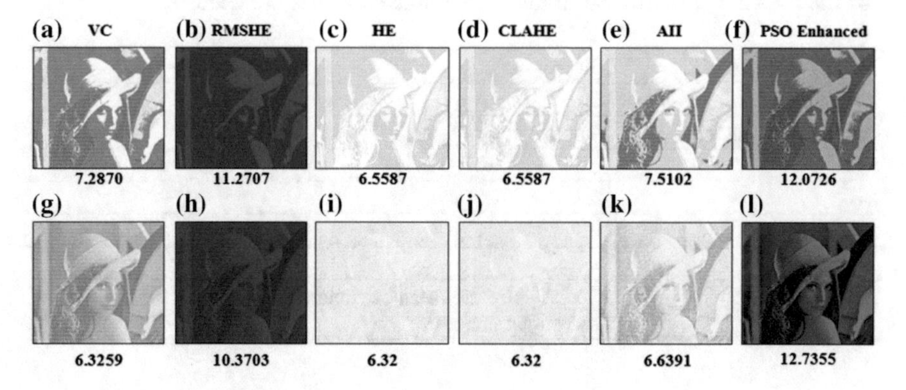

Fig. 6 Enhancement of VC shares of Lenna image

Step 3: Consider shares of the red channel. Initialize particles with random position and velocity vector

Step 4: Loop until maximum iteration

Step 4.1: Repeat for all particles

Step 4.1.1: Evaluate the difference in mean intensities (p) between $FSDI_r$ and SI_r

Step 4.1.2: If $p < pbest$, then $pbest = p$

Step 4.1.3: Go to step 4.1

Step 4.2: Best of pbest is stored as gbest and min and max value of pixel in $FSDI_r$ is stored along with *gbest*

Step 4.3: Update particles velocity using Eq. (7) and position using Eq. (8) respectively

Step 5: Go to step 4

Step 6: The value of *gbest* gives optimal solution and *min* and *max* value is the pixel value for red channel and new red shares $FSDI_{psor1}$ and $FSDI_{psor2}$ created with new pixel values

Step 7: Step 3 to Step 6 are repeated for shares of green and blue channels.

Step 8: [$FSDI_{psor1}$, $FSDI_{psog1}$, $FSDI_{psob1}$] merged to form secret VC share1 and [$FSDI_{psor2}$, $FSDI_{psog2}$, $FSDI_{psob2}$] combined to Form secret VC share2 and sent across the network

Step 9: END

The above algorithm is also used to create enhanced VC shares for original SI by following steps 2 to 9.

3.3 PSO Parameters Used

For the above algorithm, the following parameter values are used:

No. of iterations	50
No. of particles	20
Input image channels	Original color SI and VC shares of RGB color
c_1	0.1
c_2	0.1
Initial velocity	0
W^t	1
Fitness function	To minimize the mean brightness error between the input and output images

4 Results and Discussion

The proposed optimized method has been tested using Java and Matlab on customary color and gray images of size 512 × 512, such as Baboon, Lenna, Peppers, Barbara and Cameraman. The decrypted VC image is enhanced with the contemporary enhancement techniques like Histogram Equalization (HE), Contrast-limited Adaptive Histogram Equalization (CLAHE), Recursive Mean Separate Histogram Equalization (RMSHE) and Adjust Image Intensity (AII) using Matlab and their performances are compared with proposed PSO enhancement technique. The functioning of all these methods is assessed qualitatively in terms of human visual perception and quantitatively using standard metrics like Discrete Entropy, Contrast Improvement Index (CII), Histogram, Peak Signal-to-Noise Ratio (PSNR), and structured quality index (Q) to authenticate the superiority of the decrypted image.

The vital characteristic of VC is that the image includes only two intensity values (minimum and maximum) for each color channel. As discussed in Sect. 2.2, VC and FSDI VC images consist of the only two intensities 0 and 255 for each color channel. The main objective of this proposed method is to retain the significance of VC and at the same time to improve the quality of the decrypted image by adjusting the minimum and maximum color value. PSO is used to find the optimal minimum and maximum color value for each color channel that will best represent the original image.

In VC the secret message is retrieved by just superimposing or performing *OR* operation on the VC Shares and HVS discloses the secret. The result of image enhancement of VC shares of Original SI and VC shares of FSDI- half toned SI in terms of human visual perception is shown in Fig. 7a, b respectively.

The result of PSO search of 50 iterations with the gbest value obtained in each iteration and the corresponding maximum and minimum red color value calculated by the algorithm for different FSDI VC images are shown in the graph in Fig. 8a–c. The fitness value is used to minimize the mean brightness error between the input and output images. The gbest value gives the minimum difference in the mean brightness among the 20 swarms in 50 iterations.

The advantage of VC is exploited in this proposed method to hide the secret message in the form of an image by creating VC shares as shown in Fig. 4b, c and sent across the network. The shares thus generated are meaningless and look like random dots. The receiver decrypts the SI by overlapping the VC shares or by executing *OR* operation. The decrypted image is shown in Figs. 6a and 9a. Figures 6b–e and 9b–e show the result of image enhancement of decrypted image by various histogram equalization techniques and its corresponding PSNR value. The VC shares are enhanced by PSO technique and decrypted image of PSO enhanced VC shares is shown in Figs. 6f, 7a, and 9f. Visual results of PSO enhanced Image is superior to those of other HE techniques.

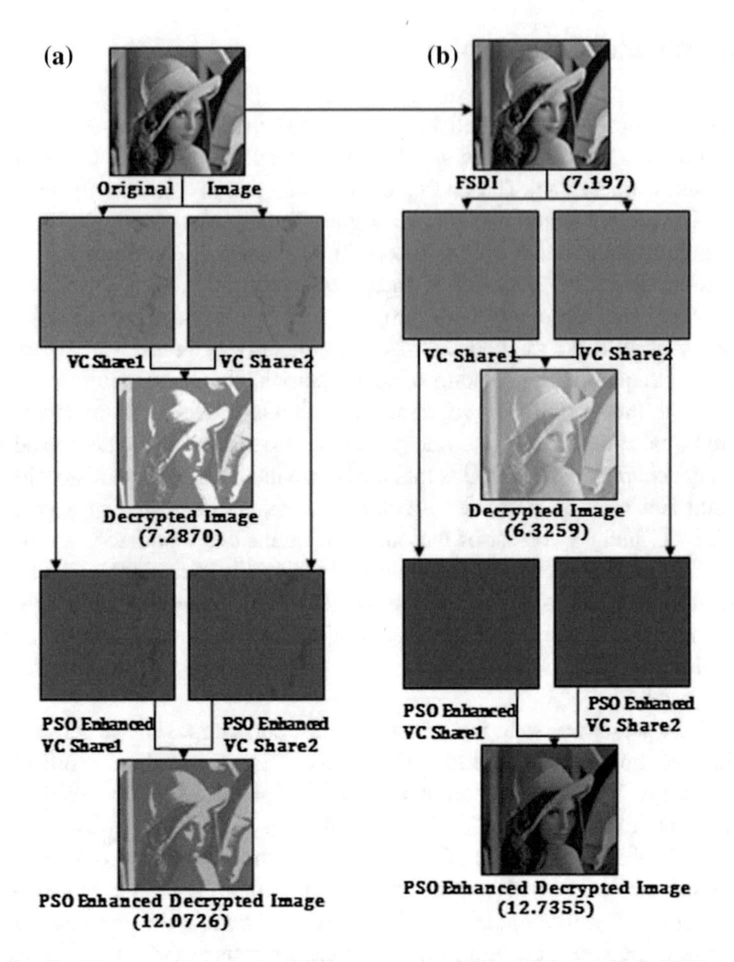

Fig. 7 Result of image enhancement of VC shares using PSO, **a** Original SI, **b** FSDI

The secret image is converted to Floyd−Steinberg's Diffusion half toned image (FSDI) and VC shares are created as shown in Fig. 7b. The decrypted SI obtained by overlapping FSDI VC shares is shown in Figs. 6g and 9g. Figures 6h−k and 9h −k show the result of image enhancement of decrypted FSDI image by various histogram equalization techniques and its corresponding PSNR value. The FSDI VC shares are enhanced by PSO technique and decrypted SI of PSO enhanced FSDI VC shares are shown in Figs. 6l, 7b and 9l. Visual perception reveals that Figs. 6l and 9l is almost similar to the original image. The encircled portions in Fig. 9a, f, g, l shows the quality, clarity and depth of PSO enhanced image compared to the original VC images.

Fig. 8 PSO search for optimization **a** Baboon **b** Lenna **c** Peppers

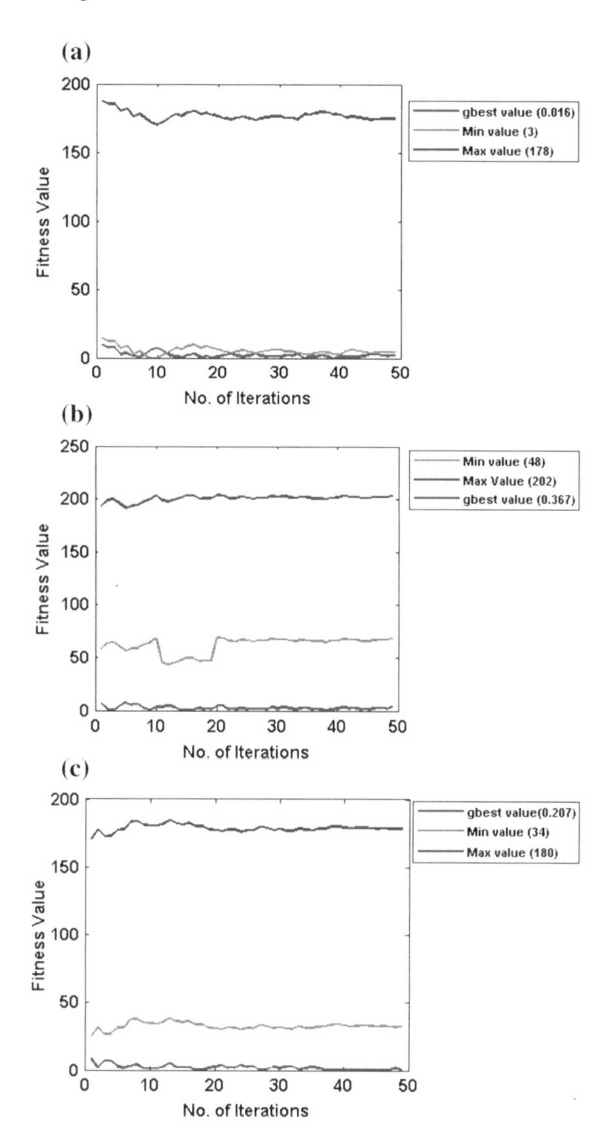

4.1 Average Information Contents (AIC)

E = entropy (I) returns E, a scalar value representing the entropy of grayscale image, where a higher value of Entropy signifies richness of the information in the output image. Entropy is a statistical measure of randomness that can be used to characterize the texture of the input image [16]. The self-information represents the number of bits of information contained in it and the number of bits we should use to encode that message. Larger entropies represent larger average information.

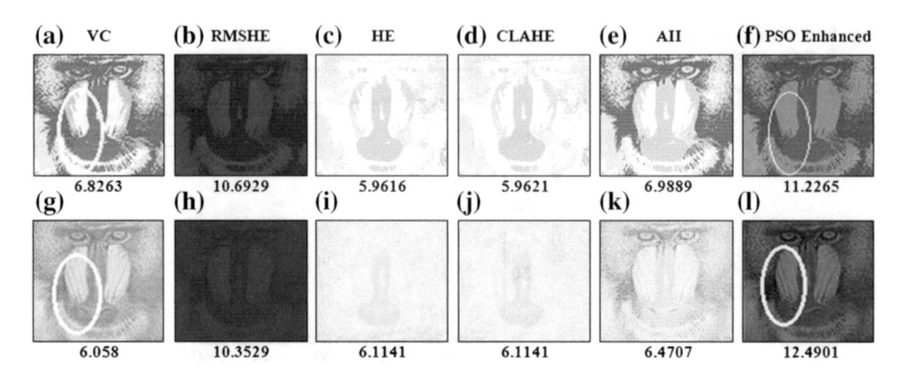

Fig. 9 Enhancement of VC shares of Baboon image

Entropy is defined as:

$$\text{AIC}(\text{Entropy}) = -\sum_{k=0}^{L-1} P(k)\log P(k) \tag{9}$$

where P(k) is the probability density function of the kth gray level.

Higher value of the AIC indicates that more information is brought out from the images. A full grayscale image has high entropy, a threshold binary image has low entropy and a single-valued image has zero entropy [17].

Table 1 presents the entropy values of different image enhancement techniques and the proposed PSO methods on the VC image. The richness of details in image is good in PSO than other techniques and is shown in Fig. 10. The entropy value of cameraman is 0.8862 and is lowest in comparison to other images. This is attributed to the fact that the image has more background than other standard images.

4.2 Contrast Improvement Index (CII)

The Contrast Improvement Index (CII) is used for evaluation of performance analysis of the proposed PSO based enhancement algorithm and is defined by

Table 1 Comparison of entropy values

Image index	Original	RMSHE	HE	CLAHE	AII	Proposed PSO	Proposed PSO (FSDI)
Lenna	0.8615	0.8615	0.9797	0.9805	1.4197	2.4050	2.4272
Baboon	0.8510	0.8510	0.7749	0.7766	1.3930	2.4288	2.4964
Peppers	0.8713	0.8713	1.1439	1.1451	1.5792	2.3797	2.3798
Barbara	0.8893	0.8893	0.8893	0.8894	0.8893	2.4742	2.5350
Cameraman	0.7308	0.7308	0.7308	0.7641	0.7308	1.7854	0.8862
Average	0.8615	0.8615	0.9037	0.9111	1.2024	2.2946	2.145

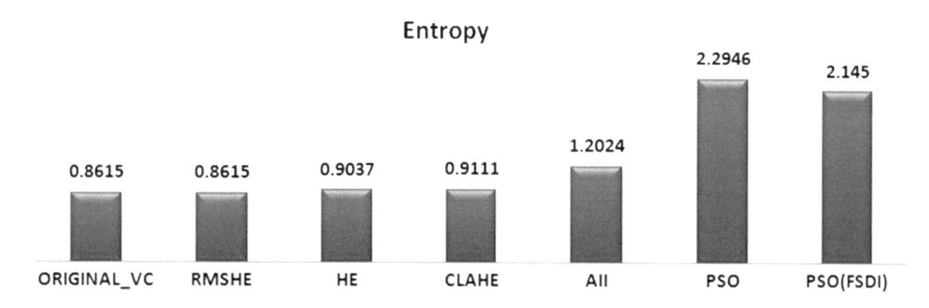

Fig. 10 Entropy chart for different enhancement techniques

$$CII = C_p/C_o \tag{10}$$

where C_p and C_o are the contrasts for the proposed and original images respectively [18]. Using the tested images a comparative study has been made and the result is shown in Tables 2 and 3. The result confirms that the proposed algorithm performs very well, and obtained results are enhanced and clearer than the original one.

Table 2 gives the comparison of CII values of images enhanced from original VC image using various enhancement techniques. Similarly Table 3 gives CII values of images enhanced from FSDI VC image using various enhancement techniques. The last column in the above two table shows that there is sharp increase in the CII values for the PSO enhanced images when compared to other

Table 2 Comparison of CII values w.r.to original VC

Image index	RMSHE	HE	CLAHE	AII	Proposed PSO
Lenna	0.298	0.3544	0.3545	0.6353	0.9395
Baboon	0.2549	0.2760	0.2761	0.4222	0.6441
Peppers	0.1804	0.4729	0.4730	0.4937	0.8276
Barbara	0.2314	0.3825	0.4465	0.5774	0.5656
Cameraman	0.2667	0.4574	0.5683	0.5774	0.5315
Average	**0.24628**	**0.3886**	**0.4237**	**0.5412**	**0.7017**

Table 3 Comparison of CII values w.r.to FSDI VC

Image index	RMSHE	HE	CLAHE	AII	Proposed PSO
Lenna	0.2863	0.3680	0.3680	0.6266	1.3405
Baboon	0.2588	0.3858	0.3858	0.6680	0.9544
Peppers	0.2667	0.3969	0.3969	0.7163	1.0875
Barbara	0.2392	0.3915	0.2660	0.4938	0.6195
Cameraman	0.2431	0.4931	0.2679	0.4935	0.6078
Average	0.2588	0.4071	0.3369	0.5996	0.9219

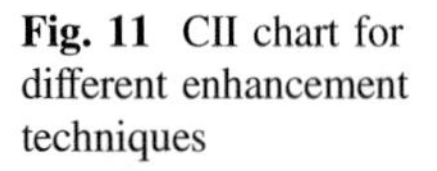

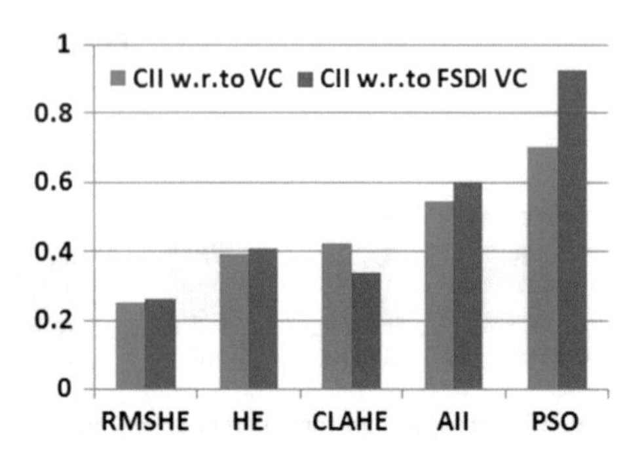

Fig. 11 CII chart for different enhancement techniques

techniques. Further, the CII value shows a sharp enhancement when FDSI VC image is enhanced using PSO and is shown in Fig. 11.

4.3 *Histogram*

An image histogram is a graphical representation of the number of pixels in an image as a function of their intensity. An Image histogram is an important tool for inspecting images. Technically, the histogram maps Luminance, which is defined by the way the human eye, perceives the brightness of different colors [19].

Every pixel in the Gray or Color image computes to a Luminance value between 0 and 255. The Histogram graphs the pixel count of every possible value of Luminance or brightness. The total tonal range of a pixel's 8 bit tone value is 0... 255, where 0 is the blackest black at the left end, and 255 is the whitest white (or RGB color) at the right end.

The height of each vertical bar in the histogram simply shows how many image pixels have luminance value of 0, and how many pixels have a luminance value 1, and 2, and 3, etc., all the way to 255 at the right end. The function imhist(I) calculates the histogram for the image I and displays a plot of the histogram [4].

The histogram comparison of red channel for various baboon images of Fig. 9 is given in Fig. 12. The vertical axis represents the number of pixels in a particular color, whereas, the variations in color is represented by the horizontal axis. The right side of the horizontal axis represents the color pixels and the left side represents black pixels. The color distribution of the Original VC and FSDI VC indicates that the pixel values are either 0 or 255. The proposed PSO optimization technique determines the best lower and higher value for pixels for RGB colors and is shown in the Fig. 12f, l.

The results confirm that PSO enhanced VC shares of Baboon image (Fig. 9f, l) consists of pixels with only two intensity values for each color channel, thus maintaining the property of VC as shown in the histogram (Fig. 12f, l).

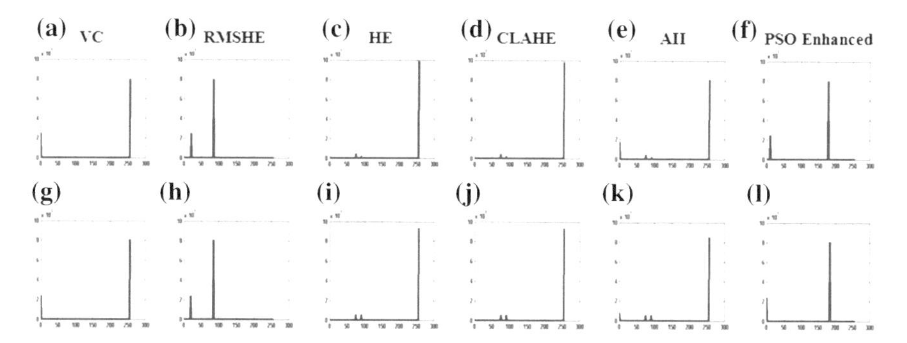

Fig. 12 Histogram comparison of various enhanced decrypted Baboon images

Table 4 Comparison of structured quality index with original VC

Image	RMSHE	HE	CLAHE	AII	Proposed PSO	Proposed PSO (FSDI)
Lenna	0.372	0.533	0.533	0.659	0.9483	0.7865
Baboon	0.350	0.459	0.459	0.827	0.9006	0.7872
Peppers	0.256	0.616	0.616	0.667	0.7832	0.7568
Barbara	0.375	0.936	0.9998	1	0.8288	0.7698
Cameraman	0.459	0.980	0.9998	1	0.8115	0.8067

4.4 *Structured Similarity Index (Q)*

The Universal Image Quality Index (Q) is a quality assessment measure for images, proposed by Wang et al. [20], and is defined as

$$Q = \frac{4\sigma_{xy}\mu_x\mu_y}{\left(\sigma_x^2 + \sigma_y^2\right)\left(\mu_x^2 + \mu_y^2\right)} \tag{11}$$

where μ_x and μ_y, σ_x and σ_y represent the mean and standard deviation of the pixels in the original image (x) and the reconstructed image (y) respectively. σ_{xy} represents the correlation between the original and the reconstructed images. The dynamic range of Q is (−1, 1) [21].

This index models any distortion as a combination of three different factors—loss of correlation, luminance distortion and contrast distortion. This quality index performs significantly better than the widely used distortion metric mean squared error. If there is no distortion q has a value 1. The proposed PSO enhanced images have higher value of q compared to other methods and is shown in Table 4.

5 Conclusion

In this paper, a novel secret image sharing scheme with image enhancement of VC shares using PSO is proposed. The loss of resolution, due to pixel expansion of VC shares is minimized using Particle Swarm Optimization technique. Results of the proposed technique are compared with other recent HE image enhancement techniques. The performance of all these methods are assessed qualitatively in terms of human visual perception and quantitatively using standard metrics like AIC, CII, Histogram, PSNR and Q, to confirm the quality of the decrypted image. The qualitative evaluation shows that there is a sharp improvement in the quality and depth of the image enhanced using PSO technique. Furthermore, the quantitative measurement values reveal that the resolution and contrast of the image are enhanced multifold in this proposed technique.

The proposed method has many advantages.

- The fitness function can be very simple and can be applied to optimize problems of large dimensions producing quality solutions more quickly.
- Tuning of input parameters and experimenting with various versions of PSO method may produce a different useful outcome.
- As PSO enhancement is done to VC Shares before sending them across the network, therefore there is no computation involved at the receiving end.
- Particle swarm optimization can be tried with different objective functions to improve the enhancement quality of color in different channels.

The proposed algorithm guarantees highly safe, secure, quick and quality transmission of the secret image with no mathematical operation needed to reveal the secret. Can be implemented in a number of applications in almost all fields like Remote sensing, Defense, Telemedicine, Agriculture, Forensics, etc. This can be used to send videos securely across the network by applying the techniques to video frames.

References

1. M. Naor, A. Shamir, A, Visual cryptography, in *Advances in Cryptology*—EUROCRYPT' 94 ed. by A. De Santis. Lecture Notes in Computer Science, vol. 950 (Springer, Berlin, 1994), pp. 1–12
2. G. Ateniese, C. Blundo, A. De Santis, R. Douglas Stinson, Visual cryptography for general access structures. Inf. Comput. **129**(2), 86–106 (1996)
3. H. Lu, Y. Li, L. Zhang, S. Serikawa, Contrast enhancement for images in turbid water. J. Opt. Soc. Am. A **32**(5), 886–893 (2015)
4. H. Lu, S. Serikawa, Underwater scene enhancement using weighted guided median filter, in *Proceedings of IEEE International Conference on Multimedia and Expo* (2014), pp. 1–6
5. Y. Wan, Q. Chen, B. Zhang, Image enhancement based on equal area dualistic sub-image histogram equalization method. IEEE Trans. Consum. Electron. **45**, 68–75 (1999)
6. S.K. Naik, C.A. Murhty, Hue-preserving color image enhancement without gamut problem. IEEE Trans. Image Process. **12**(12), 1591–1598 (2003)

7. N.M. Kwok, Q.P. Ha, D.K. Liu, G. Fang, Contrast enhancement and intensity preservation for gray-level images using multiobjective particle-swarm optimization. IEEE Trans. Autom. Sci. Eng. **6**(1), 145–155 (2009)
8. InKoo Kang, Gonzalo R. Arce, Heung-Kyu Lee, Color extended visual cryptography using error diffusion. IEEE Trans. Image Process. **20**(1), 132–145 (2011)
9. Zhi Zhou, Gonzalo R. Arce, Giovanni Di Crescenzo, Halftone visual cryptography. IEEE Trans. Image Process. **15**(8), 2441–2453 (2006)
10. P. Patil, B. Pannyagol, Visual cryptography for color images using error diffusion and pixel synchronization. Int. J. Latest Trends Eng. Technol. **1**(2), 1–10 (2012)
11. G. Ateniese, C. Blundo, A. De Santis, R. Douglas Stinson, Extended capabilities for visual cryptography. Theoret. Comput. Sci. **250**, 143–161 (2001)
12. M. Mary Shanthi Rani, G. Germine Mary, K. Rosemary Euphrasia, Multilevel multimedia security by integrating visual cryptography and steganography techniques, in *Computational Intelligence, Cyber Security and Computational Models—Proceedings of ICC3, Advances in Intelligent Systems and Computing*, vol. 412, ed. by Muthukrishnan SenthilKumar (Springer, Singapore, 2016), pp. 403–412
13. B. Padhmavathi, P. Nirmal Kumar, A novel mathematical model for (t, n)-Threshold visual cryptography scheme. Int. J. Comput. Trends Technol. **12**(3), 126–129 (2014)
14. J. Kennedy, R. Eberhart, *Swarm Intelligence* (Morgan Kaufmann Publishers, Inc., San Francisco, CA, 2001)
15. K. Gaurav, H. Bansal, Particle Swarm Optimization (PSO) technique for image enhancement. Int. J. Electron. Commun. Technol. **4**(Spl 3), 117–11 (2013)
16. R.C. Gonzalez, R.E. Woods, S.L. Eddins, *Digital Image Processing Using MATLAB'*, 2nd edn. (Gatesmark Publishing, 2009)
17. Y.-C. Chang, C.-M. Chang, A simple histogram modification scheme for contrast enhancement. IEEE Trans. Consum. Electron. **56**(2) (2010)
18. F. Zeng, I. Liu, Contrast enhancement of mammographic images using guided image filtering, in *Advances in Image and Graphics Technologies, Proceedings of Chinese Conference IGTA 2013* ed. by T. Tan, et al (Springer, China, 2013), pp. 300–306
19. S. Varnan et al., Image quality assessment techniques pn spatial domain. Int. J. Comput. Sci. Technol. **2**(3), 177–184 (2013)
20. Z. Wang et al., Image quality assessment from error visibility to structural similarity. IEEE Trans. Image Process. **13**(4), 600–602 (2004)
21. R. Kumar, M. Rattan, Analysis of various quality metrics for medical image processing. Int. J. Adv. Res. Comput. Sci. Softw. Eng. **2**(11), 137–144 (2012)

Fast Level Set Algorithm for Extraction and Evaluation of Weld Defects in Radiographic Images

Yamina Boutiche

Abstract The classification and recognition of weld defects play an important role in weld inspection. In this paper, in order to automate inspection task, we propose an aide-decision system. We believe that to obtain a satisfied defects classification result, it should be based on two kinds of information. The first one concerns the defects intensity and the second one is about its shape. The vision system contains several steps; the most important ones are segmentation and feature computation. The segmentation is assured using a powerful implicit active contour implemented via fast algorithm. The curve is represented implicitly via binary level set function. Weld defect features are computed from the segmentation result. We have computed several features; they are ranked in two categories: Geometric features (shape information) and Statistic features (intensity information). Comparative study, on synthetic image, is made to justify our choice. Encouraging results are obtained on different weld radiographic images.

Keywords Code generation • State machine • MDD • Executable UML radiographic inspection • Image segmentation • Level set • Region-based models • Features computation

1 Introduction

In industrial radiography, the more common procedure for producing a radiograph is to have a source of penetrating (ionizing) radiation (X-rays or gamma-rays) on one side of the object to be examined and a detector of the radiation (the film) on

Y. Boutiche (✉)
Research Center in Industrial Technologies CRTI, ex CSC.,
P.O. Box 64, Cheraga, 16014 Algiers, Algeria
e-mail: y.boutiche@crti.dz; bouticheyami@gmail.com

Y. Boutiche
Faculté des Sciences de L'ingénieur, Département d' Electronique,
Université Saad Dahlab de Blida, BP 120, Route de Soumaa, 09000 Blida, Algeria

H. Lu and Y. Li (eds.), *Artificial Intelligence and Computer Vision*,
Studies in Computational Intelligence 672, DOI 10.1007/978-3-319-46245-5_4

the other. This technique is very famous in NDT (Non Destructive Testing) techniques. Traditionally the human interprets the radiographic films, such task is hard and difficult when a great number of defects are to be counted and evaluated. Also, it is possible that several experts do not have the same opinion about a given film.

Nowadays, image processing is more and more introduced to automate the inspection process. Classically, the inspection vision system needs several stages, they are represented by the diagram displayed on the Fig. 1 [1, 2].

The most laborious drawback of this vision system is the necessity of a pre-processing to improve the quality of the image, otherwise the segmentation step fails, and a post processing to refine the segmentation results e.g. linking of the contour points. In this context, some works, have been published, where the segmentation steps is based on traditional techniques such as thresholding, multi-resolution approach, mathematic morphological approaches [1–5]. Such techniques require a good quality of the images. However, several preprocessing stapes are necessary such as noise remove and contrast enhancement. Unlike, the preprocessing has mismatch effects in the image.

More recently, powerful techniques are introduced in image segmentation and restoration. They are based on curve evolution theories, Partial Differential Equations PDE, and calculus of variation [6]. They are called snakes, active contours, or deformable models. The basic idea is, from an initial curve C which is given, to deform the curve until it surrounds the objects' boundaries, under some constraints from the image. The first active contour has emerged by Kass et al.'s work [7]. This work has been followed by extensive works and multiple studies in the aim to improve the capacity of extracting and segmenting images. The key elements of deformable models are: the elaboration of functionals that govern the curve evolution, the deduction of evolution equations from the functionals and finally the implementation of those equations by appropriate methods. Note that multiple choices of those keys are allowed: according to our wish to use variational evolution or not, we present the curve explicitly or implicitly. Also, the fidelity to data term of the functional is based on contour (edge-based models) [8–12] or based on statistical informations (region-based models) [13–20].

Our general objective is to develop algorithms which are able to segment, to restore, to evaluate, and to compute features of defects in radiographic images with high accuracy as much as possible and in less CPU time. Our vision to achieve such goal is: firstly, avoiding completely the preprocessing stage, especially those methods that affect the boundary localisation. For that, we have used a region-based

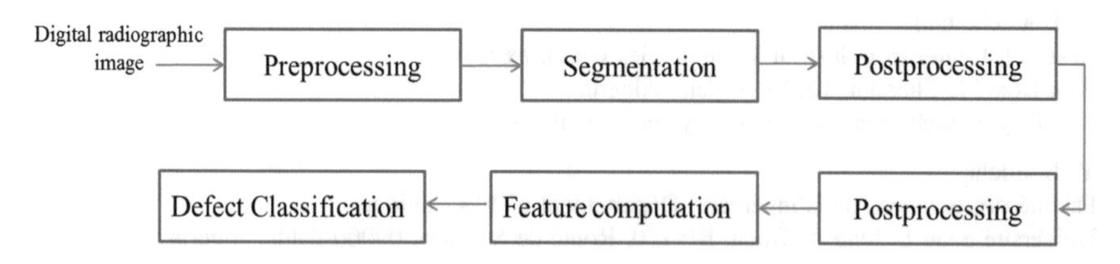

Fig. 1 The main stages of the classical vision system applied to defects' classification process

deformable model that is robust to the noise. It detects objects whose boundaries cannot be defined or are badly defined through the gradient and automatically detects interior contours. In addition, the initialization could be anywhere on the image domain not necessary surrounding the objects and it has better tendency to compute a global minimum of the functional. Secondly, we have used an implicit representation of curve to avoid any post-processing (refine the extracted contour, e.g. linking). The implicit representation is assured, in our work, by the Binary Level Set BLS function. However the extracted region (the binary object domain) is directly characterized by the final BLS, without any banalization step as the previous works do.

In this paper, we adopt an implicit region-based model named Piecewise Constant approximation PC [14], minimized via a fast algorithm. To summarize, the segmentation by such model allows us to benefit from the following main points:

- It presents a less complex design pattern for state machine implementation;
- Resolve two problems simultaneously: segmentation and restoration;
- The thickness of the extracted contour is one pixel;
- Exact location (no de-localization) of the extracted contour;
- Great compactness of the extracted contour (connected pixel);
- Two based descriptors are given at the end of the segmentation process (area and perimeter);
- Less time consuming, this is often required for industrial applications.

The rest of the paper is structured as follows: in Sect. 2 we discuss the first step of treatment which is selection of the region of interest. In Sect. 3, we focus on the segmentation algorithms and we detail the PC model and its fast minimization algorithm. In Sect. 4 we introduce the feature computation step. The experiment results on synthetic and radiographic images with some discussions are the objective of Sect. 5. Section 6 concludes the paper.

2 Selection of the Region of Interest ROI

More often, rough weld radiographic images are characterized by great dimension, very complex background, noisy and low contrast. Furthermore, those images generally contain some information about the material and its location, such information are needless for the weld's features computation stage. However, the step of selecting a region of interest (ROI) from the rough image is necessary; such selection allows to:

- Reduce time of computation;
- Avoid the processing of complex background which can be the cause of a failure contours extraction of the weld defect;
- Deduct some features directly from the final segmentation outcomes without the need to any supplementary processing.

3 Segmentation

In the chain of processing, the segmentation is the primordial step because the outcomes of this stage strongly govern the results of the next stage (features computation). However, the choice of an adequate model is necessary. A skim through the literatures, in this context, allows us to note that almost all proposed methods need a supplementary post-processing to refine the segmentation results and a step of binarization to compute the features of the segmented weld defects. For example, Wang et al. [24] propose an adaptive wavelet thresholding to extract the weld defects. Unluckily the outcomes of segmentation have to be an input of another bloc of treatments to calculate the weld features. For a state of the art about what has been done during this last decade, we refer the reader to [21] and to [22] for comparative studies.

As we have mentioned in the introduction, segmentation by using implicit active contour avoids both of the supplementary treatments. No need to refine segmentation results because it is defined in the grid image; however the extracted contour has the accuracy of a pixel. Furthermore, the area, the perimeter, and the binary object domain of the extracted object (defect) are done once the segmentation is achieved.

3.1 *Implicit Active Contour "Level Set"*

The level set method evolves a contour (in two dimensions) or a surface (in three dimensions) implicitly by manipulating a higher dimensional function, called level set $\Phi(\mathrm{x}, t)$. The evolving contour or surface can be extracted from the zero level set $C(\mathrm{x}, t) = \Phi(\mathrm{x}, t) = 0$. The great advantage of using this method is the possibility to manage automatically the topology changes of curve in evolution. However the curve C can be divided into two or more curves, inversely, several curves may merge and become a single curve [23]. A level set function can be defined as Signed Distance Function SDF (Eq. 1), its 3D graphic representation is on Fig. 2b. It can be also represented via a binary function BLS (Eq. 2), that corresponding to Fig. 2c.

$$\Phi(\mathrm{x}, t) = \begin{cases} +dist(x, C(\mathrm{t}) & if\, x \in \Omega^{+}(t) \\ -dist(x, C(\mathrm{t}) & if\, x \in \Omega^{-}(t) \end{cases}, \tag{1}$$

$$\Phi(\mathrm{x}, t) = \begin{cases} +\rho\, if\, x \in \Omega^{+}(t) \\ -\rho\, if\, x \in \Omega^{-}(t) \end{cases}, \tag{2}$$

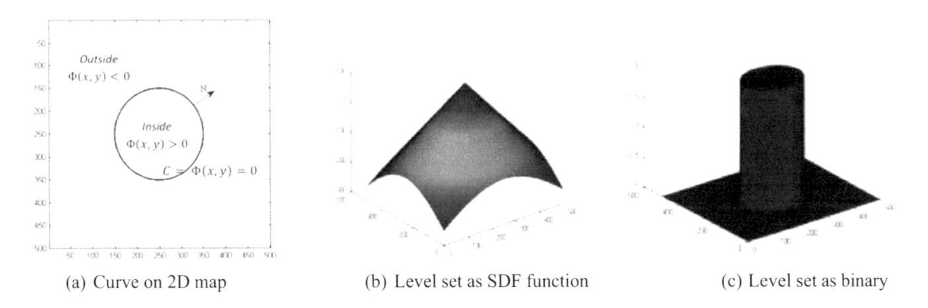

(a) Curve on 2D map (b) Level set as SDF function (c) Level set as binary

Fig. 2 The two ways to represent implicitly a curve on Fig. 2a

3.2 Piecewise Constant Approximation PC of Chan-Vese

The PC model was proposed by Chan and Vese [14]. The first PC model is based on simplifying Mumford-Shah functional [13] by approximating the resulting image u to a set of constants (two constants). The functional to be minimized is given by Eq. (3):

$$E^{PC}(c_1, c_2, C) = \lambda_1 \int\limits_{inside(C)} (u_0 - c_1)^2\, dx + \lambda_2 \int\limits_{outside(C)} (u_0 - c_2)^2\, dx + \nu|C|, \quad (3)$$

where $\nu \geq 0$, $\lambda_1 > 0$, $\lambda_2 > 0$, c_1 and c_2 are, respectively, the average image intensity inside and outside the curve. We should mention that, according to the Hausdorff measure in 2D, the last term in the functional (3) represents the length of the contour. However, in the convergence, this term is only the perimeter of the extracted object. The Eq. (3) can be written via an implicit representation of the curve using level set function Φ introduced by [23] as follows:

$$F^{PC}(c_1, c_2, \Phi) = \lambda_1 \int\limits_{\Omega} (u_0 - c_1)^2 \mathrm{H}_\varepsilon(\Phi(x))dx + \lambda_2 \int\limits_{\Omega} (u_0 - c_2)^2 (1 - \mathrm{H}_\varepsilon(\Phi(x)))dx + \nu \int\limits_{\Omega} |\nabla \mathrm{H}_\varepsilon(\Phi(x))dx| \quad (4)$$

where Φ is the level set function and $\mathrm{H}_\varepsilon(\Phi(x))$ is the regularized version of Heaviside function, used to identify the inside and outside regions. It is formulated by Eq. (5) and its derivative δ_ε by Eq. (6).

$$\mathrm{H}_\varepsilon(\Phi(x)) = \frac{1}{2}\left[1 + \frac{2}{\pi}\arctan\left(\frac{z}{\varepsilon}\right)\right], \quad (5)$$

$$\delta_\varepsilon(x) = \frac{1}{\pi}\frac{\varepsilon}{\varepsilon^2 + z^2}, \ z \in \mathcal{R} \quad (6)$$

Keeping Φ fixed and minimizing the functional F^{PC} with respect to c_1 and c_2 we get:

$$c_1 = \frac{\int_\Omega u_0(\mathrm{x})\mathrm{H}_\varepsilon(\Phi(\mathrm{x}))d\mathrm{x}}{\int_\Omega \mathrm{H}_\varepsilon(\Phi(\mathrm{x}))d\mathrm{x}}, \tag{7}$$

$$c_2 = \frac{\int_\Omega u_0(\mathrm{x})(1-\mathrm{H}_\varepsilon(\Phi(\mathrm{x})))d\mathrm{x}}{\int_\Omega (1-\mathrm{H}_\varepsilon(\Phi(\mathrm{x})))d\mathrm{x}}. \tag{8}$$

For c_1 and c_2 fixed, the according Euler-Lagrange equation that allows the evolution of the curve is given by the following Eq. (9)

$$\frac{\partial \Phi}{\partial t} = \delta_\varepsilon(\Phi)\left[\nu\, div\left(\frac{\nabla \Phi}{|\nabla \Phi|}\right) - \lambda_1(u_0 - c_1)^2 + \lambda_2(u_0 - c_2)^2\right]. \tag{9}$$

When the minimization of the functional is reached, the result image u can be represented by the following formulation:

$$u^{PC}(\mathrm{x}) = c_1\mathrm{H}_\varepsilon(\Phi(\mathrm{x})) + c_2(1-\mathrm{H}_\varepsilon(\Phi(\mathrm{x}))).$$

To reach the minimum of the energy, we have to solve the PDE (9) several times until the energy being stationary (principle of gradient descent method). In the next subsection, we introduce a faster and more stable algorithm to minimise such functional.

3.3 Fast Algorithm to Minimized PC Model

In almost all active contours, the minimization of the energy is assured by the gradient descent method GD. It is based on introducing a virtual temporal variable in the corresponding Euler-Lagrange static energy (resolve partial differential equation) and evolves it iteratively until it reaches the minimum. The Courant *Friedrichs Lewy (CFL)* condition is required to insure a stable evolution. As a consequence, the time step must be very small and the evolution becomes time-cost consuming.

B. Song et al. [24] proposed algorithm able to minimize the functional without needing to solve any PDE and consequently no numerical stability conditions are required. Instead of that, we sweep all level set points, and we test each point to check if the energy decreases or not when we change a point from the inside of the curve (level set) to the outside. The principle of this algorithm is applied on the PC Chan-Vese model presented above. The main steps of the algorithm are given as follows:

Algorithm: Sweeping algorithm applied on PC model
• Give any initial partition of the image domain, Set $\Phi = 1$ for one part and $\Phi = -1$ for the another one. • Compute c_1 and c_2 that are the average image intensities for $\Phi = 1$ and $\Phi = -1$ respectively. • Compute m, n that are the area (number of pixels) for $\Phi = 1$ and $\Phi = -1$ - **If** $\Phi(x) = +1$, then compute the difference between the new and old energy: $$\Delta F_{12} = \left[(u_0 - c_2)^2 \frac{n}{n+1} + \nu P\right] - \left[(u_0 - c_1)^2 \frac{m}{m-1} + \nu P\right]$$ - **If** $\Delta F_{12} < 0$, then change $\Phi(x)$ from $+1$ to -1. - **If** $\Phi(x) = -1$, then compute the difference between the new and old energy: $$\Delta F_{21} = \left[(u_0 - c_1)^2 \frac{m}{m+1} + \nu P\right] - \left[(u_0 - c_2)^2 \frac{n}{n-1} + \nu P\right]$$ - **If** $\Delta F_{21} < 0$, then change $\Phi(x)$ from -1 to $+1$. • Repeat the step 2 until the total energy F remains unchanged.

Remind that ν is a positive constant to penalise the minimum length of evolving curve which is analytically given by the last integral in the Eq. (4). It can be approximated numerically as follows:

$$P = \sum_{(i,j)} \sqrt{\left(\mathrm{H}_\varepsilon(\Phi_{i+1,j}) - \mathrm{H}_\varepsilon(\Phi_{i,j})\right)^2 + \left(\mathrm{H}_\varepsilon(\Phi_{i,j+1}) - \mathrm{H}_\varepsilon(\Phi_{i,j})\right)^2}.$$

On Fig. 1 we represent detailed diagram of the above algorithm (Fig. 3).

3.4 *Why PC Model via Sweeping Minimization?*

In this subsection, we are going to justify our choice of the PC model among other models. Generally speaking, all *region-based* models could be ranged in three classes as follows: *Global region-based models*, Local *region-based* models, and *Global Local (hybrid) region-based* models. More often the two last families use the kernel Gaussian function to get the local propriety.

The first row of Fig. 4 shows the segmentation results of the local binary fitting model LBF [15] and PC models. We have made a zoom on the same location for both outcomes. It is clear to show that the LBF model extracts contour with some delocalisation because of using the Gaussian kernel function. The second row of Fig. 4 represents the corresponding mesh of the final level set function for the LBF model (left hand) and PC model (right hand). We display, on the third row, the

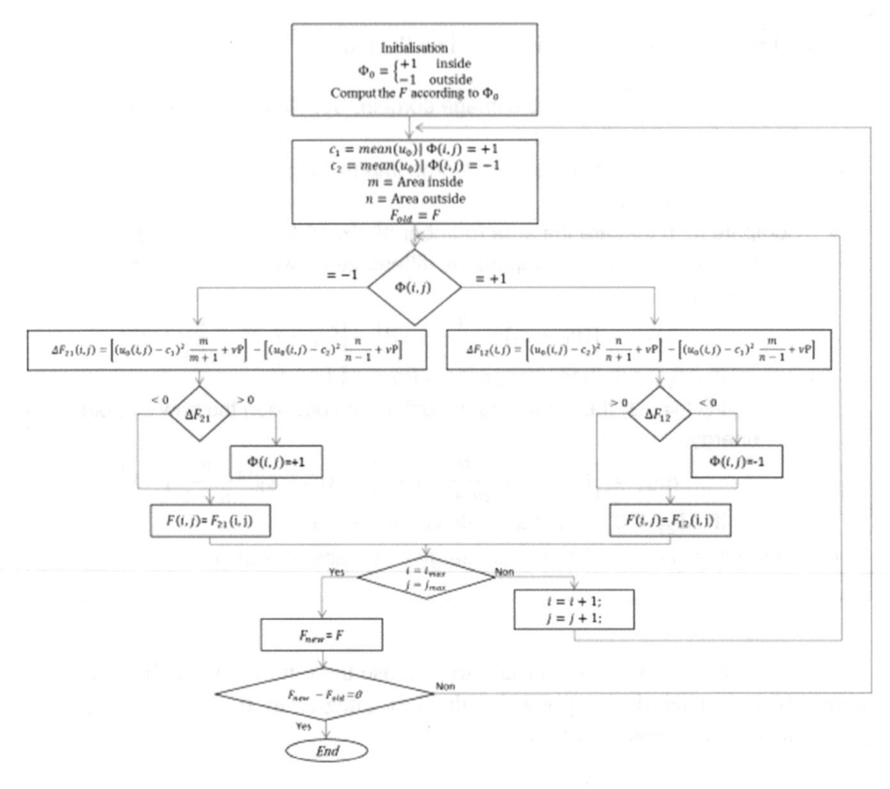

Fig. 3 Diagram of the fast algorithm minimizing the PC model

profile lines of final level set of both models. This experiment shows that the best and high accurate segmentation result is obtained by PC model.

From another point of view, the minimisation functional via gradient descent method needs a stooping criteria, which is generally defined as the mean square error MSE or the relative mean square error RMSE between the current level set and the previous one, that must be less than a given small constant ϵ. In our case, this constant is automatically taken to 0 ($F_{new} = F_{old}$).

More often, for many industrial applications, a compromise between the quality of treatments and time consuming is required. For this reason we have chosen the PC model minimised via the sweeping principle which allows very fast convergence and great accuracy.

The weak point of PC model is its less ability to deal with inhomogeneous intensity distribution. Such drawback does not matter in our context, because we treat small selected region.

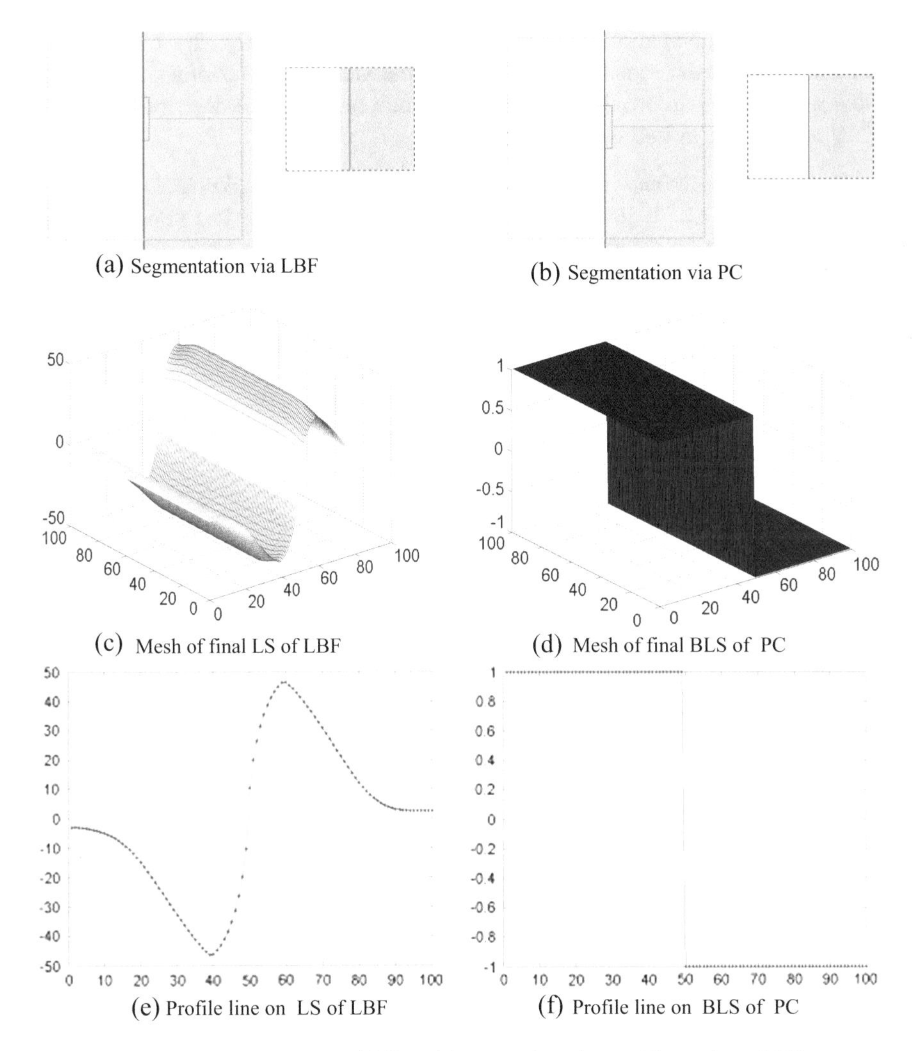

(a) Segmentation via LBF
(b) Segmentation via PC
(c) Mesh of final LS of LBF
(d) Mesh of final BLS of PC
(e) Profile line on LS of LBF
(f) Profile line on BLS of PC

Fig. 4 Comparison between LBF and PC models on synthetic image. First column LBF outcomes with $\sigma = 3$, 350 iterations $CPU = 15.94$ s. Second column PC outcomes, total sweep = 3, $CPU = 0.001$ s

4 Features Computation

The computation of weld defect features is very helpful for their classification and recognition. For example, the value of roundness called also compactness that is included in [**0, 1**] gives an idea about the kind of defect. If it has little value, so the defect has sharp shape, consequently it is a crack or a lack of fusion. Contrary if roundness has large value (near 1), the defect has round shape such as porosity or inclusion.

Once the segmentation is achieved, we have got directly: the binary object domain which is the same as the final BLS, the area and the perimeter of defects. Now we are going to use those outcomes to compute the more descriptive features. They are classed in two classes as follows:

- *Geometric Descriptions GeoD*: such class of descriptors allows the description of the shape and region of a given segmented image. In this paper, we have calculated the boundary descriptions BD and region descriptors RD that characterised as well as possible each class of defects.
- *Statistic Descriptions StatD*: The reflection of defects on radiographic image change according to its kind. For example, the lack of penetration is reflected by a dark region (weak intensity) on the welded joint. On the contrary, inclusion of metal defects are reflected by clear regions (high intensity).

4.1 Geometric Features

Geometric features discriminate well the weld defects. Meanwhile, they will be useful for their classification and recognition. From large descriptors of shape and region that have been introduced in [25, 26], we select some of them that seem to well discriminate the weld defects. Remember that the perimeter and area have already been getting at the end of the segmentation process.

4.1.1 Boundary Descriptions

- *Basic boundary descriptors*: the perimeter, length, width, basic rectangle (the smallest rectangle that contains every point of the object), are used to compute more descriptive shape's features.
- *Eccentricity*: The eccentricity actually represents the ratio of the minor axis (width) to major axis (length). It value reflects how much the shape looks like an ellipse.

$$E_{cc} = \frac{W}{L} \tag{10}$$

- *Elongation*: Elongation takes its values between 0 and 1 ($Elo \in [0, 1]$), it describes how much the shape is rectangular. For circle it is zero, for the large shape its value approaches 1.

$$Elo = 1 - \frac{W}{L}, \tag{11}$$

4.1.2 Region Descriptions

- *Basic region descriptors*: The most used basic region descriptor is the area, which allows us to compute several discriminant features as follows:
- *Centroid C*: The centroid's coordinates are the arithmetic mean of the pixels coordinates of the extracted region $((X_r, Y_r) = \Phi(i, j) \in \Omega^+)$,

$$C\begin{pmatrix} \bar{x} \\ \bar{y} \end{pmatrix} = \text{mean}\begin{pmatrix} X_r \\ Y_r \end{pmatrix} \tag{12}$$

- *Roundness R:* Roundness of an object can be determined by using the formula:

$$R = \frac{4\pi A_s}{P^2}, \tag{13}$$

where A_s is the area of the shape and P its perimeter.

Remark If the Roundness is greater than 0.90, then the object is circular in shape.

- *Dispersion or Irregularity IR*: This measure defines the ratio between the radius of the maximum circle enclosing the region and the maximum circle that can be contained in the region. Thus, the measure will increase as the region spreads.

$$IR = \frac{\max\left(\sqrt{(x_i - \bar{x})^2 + (y_i - \bar{y})^2}\right)}{\min\left(\sqrt{(x_i - \bar{x})^2 + (y_i - \bar{y})^2}\right)}, \tag{14}$$

where $(\bar{x}, \bar{y})$ represent the co-ordinates of the centre of mass (Centroid) of the region.

- Rectangularity Rec: This extent gives an idea about how rectangular the shape is. It is formulated by:

$$\text{Rec} = \frac{A_s}{A_r}, \tag{15}$$

where A_r is the area of the minimum bounding rectangle.

- *Equivalent Diameter Ed*: Equivalent Diameter or Heywood Diameter is the diameter of a circle that has the same area as the defect. Its mathematical formula is given by:

$$Ed = \sqrt{\frac{4A_s}{\pi}}. \quad (16)$$

4.2 Statistical Features or Histogram Based Features

Statistic features are related to the grey scale (intensity) of the segmented region (weld). However we get back the segmented region with its original grey scale by multiplying the final binary level set by the region of interest ROI image, let's call it "GSImg". From the image (GSImg), we compute the following statistical features:

- Maximum, Minimum and Mean intensity: Scalars specifying the value of the greatest, the lowest and the mean of the intensity in the region (weld defect).
- *Variance V*:

$$V = \text{mean}\left(GSImg^2\right) - (\text{mean}(SImg))^2, \quad (17)$$

- *Standard deviation*: Standard deviation is the square root of the variance.

$$\sigma = \sqrt{V}. \quad (18)$$

- *Weighted Centroid*: Weighted Centroid is computed as the simple Centroid but taking into account the grey scale of each pixel.

5 Experiment Results

In this section, we present the outcomes of different steps discussed above on three x-rays images. For all experiments, we use dashed green line to present the contour initialization and the solid red line for final contour.

We apply the implemented algorithms on some radiographic images that contain different kinds of weld defects. Here the user should select the region of interest which is the defect. The initialisation of contour is automatically produced according to the selected region. For all following experiments, we have fixed the two parameters as follows: $\nu = 0.00045 \times 255^2$ and $\varepsilon = 0.1$.

We have chosen three radiographic images, which present different kinds and number of defects. The first experiment exhibit in the Fig. 5a, is an image containing two lack of penetration defects named *D1* and *D2*. The second one Fig. 5b contains a crack defect labelled *D3*. The third image Fig. 5c reflects three metal inclusion defects labelled *D4*, *D5* and *D6*. On each figure, we present the selected ROI on the original image. The segmentation results are presented on the Table 1. The first column represents the ROI region with the initial (green line) and final

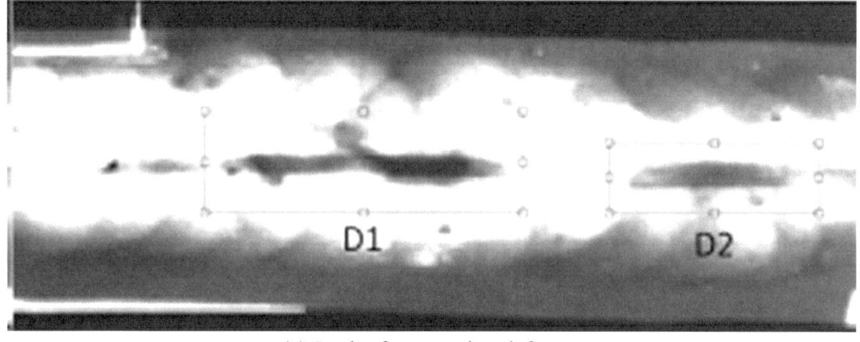

(a) Lack of penetration defects

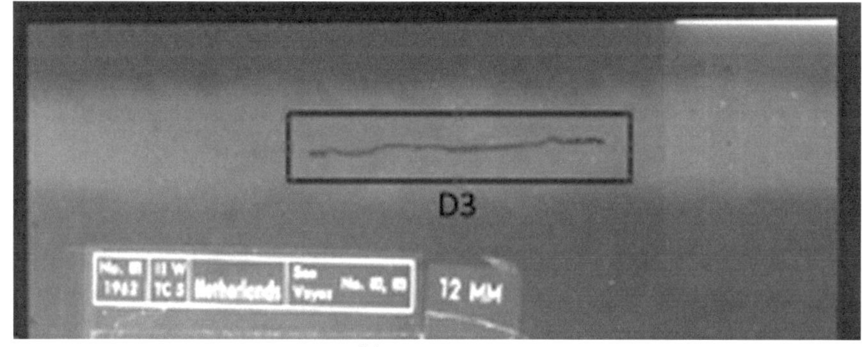

(b) Crack defect

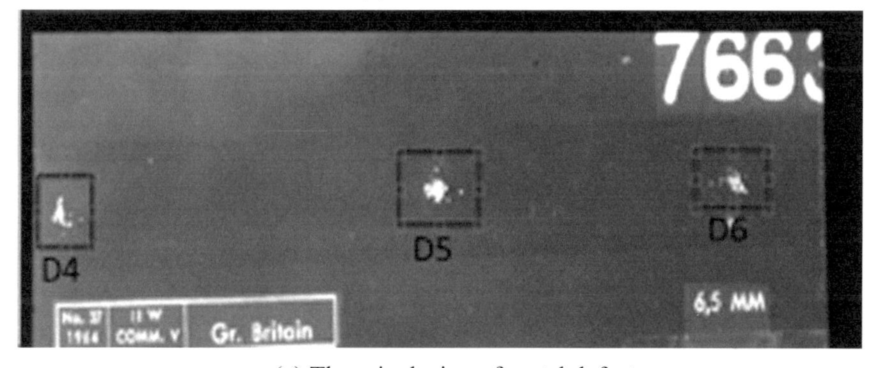

(c) Three inclusion of metal defects

Fig. 5 Radiographic images used in the present work with selection of region of interest

contour (red line). In the second one, we show the corresponding mesh of final BLS and in the last column we display the binary object domain with the smallest rectangle (cyan) and Centroid (cyan start). We summarized on Table 2 the outcomes of feature computation step.

Table 1 Experimental results of the PC model

	Initial and final contour	Mesh of final LSF	Features
D1			
D2			
D3			
D4			
D5			
D6			

Table 2 Computed features of the three above experiments, BD: Boundary Descriptors; RD: Region Descriptors; HD: Histogram Descriptors

		Exp. 1		Exp. 2	Exp. 3		
		D1	D2	D3	D4	D5	D6
BD	Perimeter	163.45	91.55	426.66	47.89	50.14	42.24
	Length	61	35	191	15	15	12
	Width	14	10	11	8	12	10
	Eccentricity	0.22	0.28	0.057	0.53	0.80	0.83
	Elongation	0.77	0.71	0.94	0.46	0.2	0.16
RD	Area	322	176	450	66	111	70
	Centroid	[36 14]	[23 10]	[108 12]	[17 31]	[34 23]	[41 22]
	Roundness	0.15	0.26	0.031	0.36	0.55	0.49
	Dispersion	3.08	3.04	2.32	1.72	2.08	1.33
	Rectangularity	0.37	0.50	0.21	0.55	0.61	0.58
	Equiv. diameter	20.24	14.96	23.93	9.16	11.88	9.44
HD	Weighted centroid	[35 13]	[23 10]	[107 12]	[17 31]	[34 23]	[41 22]
	Max intensity	165	179	92	247	255	216
	Mean intensity	96.43	125.65	75.79	203.36	209.04	169.37
	Min intensity	38	77	47	151	148	121
	Variance	1197.24	937.07	82.93	892.05	1033.5	743.91
	S. deviation	34.60	30.61	9.10	29.86	32.14	27.27

5.1 Trouble Situations

In this section, we discuss some situations in which the extraction of defects doesn't perform well or totally fails. On Fig. 6 we present such situation, where the radiographic image contains a vertical fracture, such kind of defects is reflected by a continuous, practically, similar grey scale along of the width of the welded joint. However the extraction of the defect's boundaries becomes very difficult or impossible. Consequently, the computation of features can't be done with accuracy.

5.2 The Performance of the Fast PC Model

We are going to focus on the high performance of the fast PC algorithm in terms of CPU time consuming. For that we display the number of sweeping times to extract each weld defect and the corresponding CPU time. As Table 3 shows the segmentation process doesn't take more than two percent of the second. Such rapidity is often required in industrial applications (Table 3).

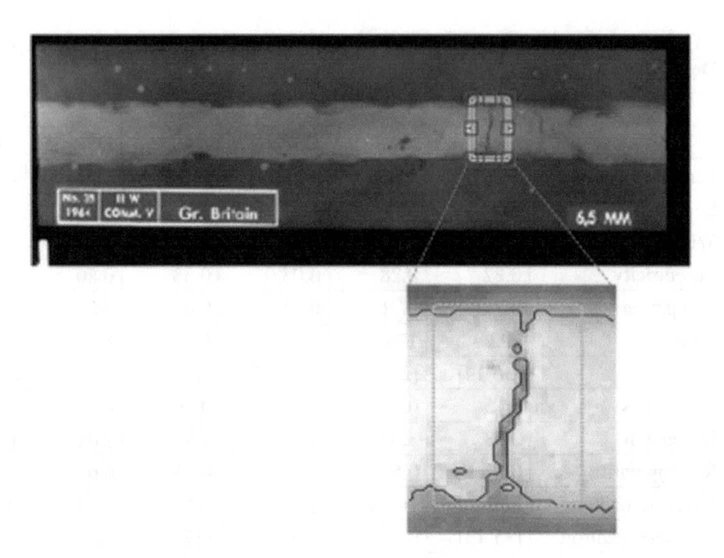

Fig. 6 Case where the extraction of defect is not achieved with good accuracy

Table 3 The performance of the fast PC model in CPU time consuming (Processor: core(TM) i7-2600 CPU 3.40GHZ, RAM: 4 Go)

	D1	D2	D3	D4	D5	D6
Size	[25 × 70]	[17 × 46]	[21 × 16]	[62 × 51]	[52 × 70]	[34 × 69]
Sweep	7	9	7	14	8	7
CPU (s)	0.01	0.01	0.01	0.02	0.01	0.01

6 Conclusion

The proposed work is a contribution to automate the radiographic inspection. Such task involves several steps of image processing and analysis.

In this paper, we have proposed and implemented algorithms that allow the extraction and feature computation of several defects' features. The first stage is the segmentation, which is primordial in the vision system. For this step, we have adopted the powerful implicit PC deformable model. Such model deals greatly with mediocre quality of radiographic images. However we have omitted the noise reduction and contrast enhancement steps that could introduce some modification on the contour's location. Furthermore, we have used an implicit representation of the contour to get perfect connected extracted contours without any supplementary steps to refine it.

From another point of view, the PC functional was minimised via a fast algorithm which does not need to control stability and CFL condition as the traditional gradient descent method required; thereby a very fast convergence is obtained.

For the features computation, we have got the two important features (perimeter and area of defect) directly from the segmentation process, as for the rest of the features, they are computed straightforward of their mathematical formulas. The final binary level set is used to get back the intensity of the defect, from which we compute the statistical features. We have also presented a difficult situation when the contour extraction couldn't be achieved correctly.

References

1. N. Nacereddine, M. Zelmat, S. Belafa, M. Tridi, Weld defect detection in industrial radiography based digital image processing, in *3rd International Conference: Sciences Of Electronic, Technologies Of Information And Telecommunications, Conference Proceedings*, 2005
2. I. Valavanis, D. Kosmopoulos, Multiclass defect detection and classification in weld radiographic images using geometric and texture features. Expert Syst. Appl. Elsevier **37**(12), 7606–7614 (2010)
3. A. Kehoe, The detection and evaluation of defects in industrial images. PhD thesis, University of Surrey (1990)
4. A. Laknanda, R. Anand, P. Kumar, Flaw detection in radiographic weld images using morphological approach. NDT & E International, Elsevier **39**(1), 29–33 (2006)
5. Y. Haniza, A. Hamzah, Y. Hafizal, Automated thresholding in radiographic image for welded joints. Nondestructive Testing and Evaluation, Taylor & Francis **27**(1), 69–80 (2012)
6. G. Aubert, P. Kornprobst, *Mathematical Problems in Image Processing Partial Differential Equations and the Calculus Of Variations* (Springer, New York, 2006)
7. M. Kass, W. Andrew, D. Terzopoulos, Snakes: Active contour models. Int. J. Comput. Vision **1**(4), 321–331 (1988)
8. V. Caselles, R. Kimmel, G. Sapiro, Geodesic active contours. Int. J. Comput. Vision **22**(1), 61–79 (1997)
9. N. Paragios, R. Deriche, Geodesic active regions and level set methods for motion estimation and tracking. Comput. Vis. Image Underst. **97**(3), 259–282 (2005)
10. A. Vasilevskiy, K. Siddiqi, Flux-maximizing geometric ows. IEEE Trans. Pattern Anal. Mach. Intell. **24**(12), 1565–1578 (2002)
11. C. Li, C. Xu, C.F. Gui, Level set evolution without re-initialization: a new variational formulation, in *2005 IEEE Computer Society Conference on Computer Vision and Pattern Recognition (CVPR'05), Conference Proceedings*, pp. 430–436, 2005
12. G.P. Zhu, S. Zhang, C. Wang, Boundary-based image segmentation using binary level set method. Opt. Eng. SPIE Lett. **46**(5), 1–3 (2007)
13. D. Mumford, J. Shah, Optimal approximation by piecewise smooth function and associated variational problems. Commun. Pure Appl. Math. **42**(5), 577–685 (1989)
14. L. Vese, T. Chan, A multiphase level set framework for image segmentation using the Mumford and shah model. Int. J. Comput. Vision **50**(3), 271–293 (2002)
15. H. Lu, S. Serikawa, Y. Li, L. Zhang, S. Yang, X. Hu, Proposal of fast implicit level set scheme for medical image segmentation using the chan and vese model. Appl. Mech. Mater. **103**, 695–699 (2012)
16. H. Lu, Y. Li, S. Nakashima, S. Yang, S. Serikawa, A fast debris flow disasters areas detection method of earthquake images in remote sensing system. Disaster Adv. **5**(4), 796–799 (2012)
17. L. Wang, L. He, A. Mishra, C. Li, Active contours driven by local Gaussian distribution fitting energy. Sig. Process. **89**(12), 2435–2447 (2009)

18. Y. Yu, W. Zhang, C. Li, Active contour method combining local fitting energy and global fitting energy dynamically, in *Medical Biometrics, Conference Proceedings*, pp. 163–172. 2010
19. Z. Kaihua, Z. Lei, S. Huihui, Z. Wengang, Active contours with selective local or global segmentation a new formulation and level set method. Image Vis. Comput. **28**, 668–676 (2010)
20. H. Wang, T. Huang, Z. Xu, Y. Wang, An active contour model and its algorithms with local and global Gaussian distribution fitting energies. Inf. Sci. Elsevier **263**, 43–59 (2014)
21. R. Romeu, D. Silva, D. Mery, "State-of-the-art of weld seam radiographic testing: Part ii pattern recogtion", www. ndt.net- Document Information. Mater. Eval. **65**, 833–838 (2007)
22. V. Rathod, R. Ananda, A. Ashok, Comparative analysis of NDE techniques with image processing. Nondestruct. Testing Eval. **27**, 305–326 (2012)
23. S. Osher, J. Sethian, Fronts propagating with curvature de-pendent speed: Algorithms based on hamilton-jacobi formulations. J. Comput. Phys. **79**, 12–49 (1988)
24. B. Song and T. Chan, "A fast algorithm for level set based optimization", CAM-UCLA, vol. 68, 2002
25. W. Pratt, *Digital Image Processing*, PIKS Inside, 3rd edn. William K. Pratt, ISBN 0-471-22132-5, 2001
26. T. Acharya, A.K. Ray, *Image Processing Principles and Applications* (Wiley, mc. Publication, 2005)

Efficient Combination of Color, Texture and Shape Descriptor, Using SLIC Segmentation for Image Retrieval

N. Chifa, A. Badri, Y. Ruichek, A. Sahel and K. Safi

Abstract In this article we present a novel method of extraction and combination descriptor to represent image. First we extract a descriptor shape (HOG) from entire image, and in second we applied method of segmentation and then we extract the color and texture descriptor from each segment in order to have a local and global aspect for each image. These characteristics will be concatenate, stored and compared to those of the image query using the Euclidean distance. The performance of this system is evaluated with a precision factor. The results experimental show a good performance.

Keywords Combined descriptor · HOG · LBP · HSV · SLIC superpixel segmentation · Retrieval image

1 Introduction

More visual information is a major consequence of convergence between computer sciences and audio-visual. More and more applications occur, use and disseminate visual data including fixed and moving images.

This evolution aroused a need for developing research techniques of multimedia information and in particular, image search. Content based image (CBIR) is new technique to overcomes problems posed by the text search, and allows improve interrupted applications in various fields. The performance of image retrieval systems depends to a much of the choice of descriptors and technical employees to extract them. A descriptor is defined as the knowledge used to characterize the information contained in the images; many descriptors are used in research systems

N. Chifa (✉) · A. Badri · A. Sahel · K. Safi
Faculty of Sciences and Techniques (FSTM), EEA&TI Laboratory,
Hassan II University of Casablanca, Mohammedia, Morocco
e-mail: nawal.chifa@gmail.com

Y. Ruichek
IRIES-SET-UTBM, 90010 Belfort Cedex, France

H. Lu and Y. Li (eds.), *Artificial Intelligence and Computer Vision*,
Studies in Computational Intelligence 672, DOI 10.1007/978-3-319-46245-5_5

to describe the pictures. we distinguish two kinds, the global one how reflecting the overall visual appearance of an image [1], such as color histogram [2, 3], edge histogram, texture co-occurrence [4], the local binary pattern (LBP) [5] and so on. However, these features tend to lose the spatial correlation among pixels. To overcome the problem, many researchers proposed the local features who focus mainly on key points [6], the most popular ones: SIFT [7], GLOH [8], SURF [9], HOG [10].

In our work, we have try to choice and combine a multiples descriptor to obtain an efficient presentation of image, in order to have good performance of our CBIR.

The rest of this paper is organized as follows. In Sect. 2 we introduce our related works. In Sect. 3 our method of extraction is detailed. In Sect. 4, we compare and evaluated the experimental results. At last section we conclude the paper.

2 Related Work

Color, texture and shape information have been the substantial features of an image in content based image retrieval, but this basic features wasn't good to describe the spatial information in image. In our work we use novel method of extraction and combined these features, first we opted for image representation based on histogram because is one of the most common features; specify color histogram [1], Local Binary Pattern (LBP) [5] and Histogram of Oriented2Gradient (HOG) [9]. And to overcome the spatial limitation of these features, we segment the image into regions, and every segment was converted to HSV space, the histogram color is extracted, for the same segment in grey level we extract the LBP descriptor and at end we concatenate the two vectors, we do that to every segment on image, and concatenate the all results vectors to obtain a local color and texture description of an image. For the shape (HOG) descriptor is applied it to the overall image because is a local descriptor; and at last we add it to the first vector (combined of histogram HSV and LBP), Fig. 1 shows an example of this method.

3 Techniques and Methods Used

Our objective in this paper is to extract the global feature locally, for that we used many techniques: superpixel Slic segmentation, HSV histogram, Histogram of uniform LBP, histogram of gradient.

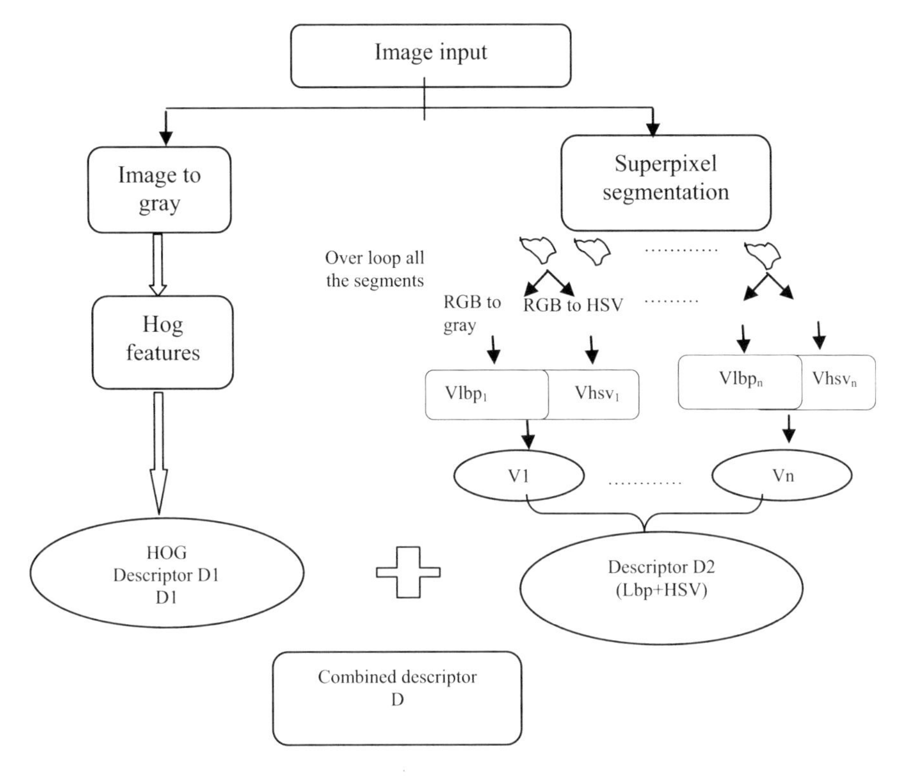

Fig. 1 Illustration of our method of features extraction and combination

3.1 SLIC Superpixel

To explore the semantic meanings of an image we used different methods of segmentation, in this word we used simple one: superpixels who segment each image into approximately (25–200) regions using Normalized Cuts [11]. This method has many desired properties [12]:

- It reduces the complexity of images from hundreds of thousands of pixels to only a few hundred superpixels.
- Each superpixel is a perceptually consistent unit, i.e. all pixels in a superpixel are most likely uniform in, say, color and texture.
- Because superpixels are results of an over segmentation, most structures in the image are conserved.

Fig. 2 Image segmented using the SLIC algorithm into superpixels

Many existing algorithms in computer vision use the pixel-grid as the method of segmentation. In this paper we have used simple the linear iterative clustering (SLIC) algorithm that performs a local clustering of pixels in the 5-D space defined by the L; a; b values of the CIELAB color space [13]. SLIC is simple to implement and easily applied in practice, the only parameter specifies the desired number of superpixels [14]. An example of SLIC segmentation image is shown in Fig. 2.

3.2 Descriptor HOG

Histogram of Oriented gradient descriptors have been introduced by Dalal and Triggs [10], the interest of this descriptor is to calculate the distribution of intensity gradients or edge directions in localized portion of an image. First they divide the image into small connected cells, and for each region compiling a histogram of gradient directions or edge orientations for the pixels within the cell. The combination of these histograms represents the local shape descriptor. Figure 3 gives an example.

In our work we extract the HOG descriptor from full image before segmented image; because the feature Hog is extract from block and give a local description for the shape information.

Fig. 3 Example for calculation the LBP operator

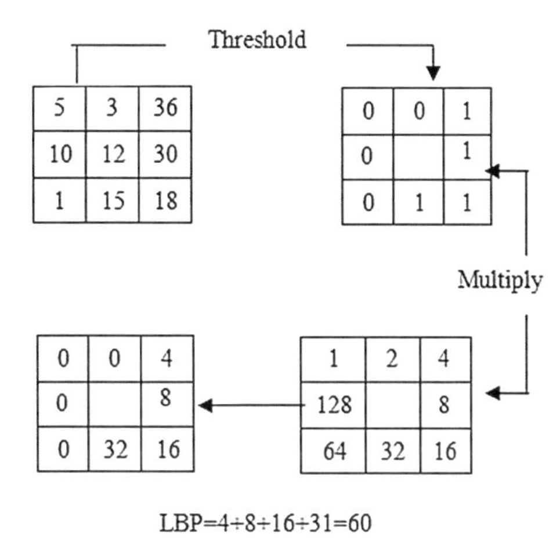

3.3 Histogram Color

Color is perhaps the most expressive of all the visual features and has been extensively studied in the image retrieval research during the last decade [15]. Swain and Ballard [2] proposed a color histogram. Then the distance between two images is calculated utilizing the histogram intersection method. This method is very simple to implement and produces a reasonable performance. However, the main disadvantage of histogram color is that it is not robust to significant appearance changes because it does not include any spatial information. This is why we use extraction from segment of image, and we combined all the histogram to have a local color descriptor.

To describe an image color, we have multiple color space RGB, HSV, YCrCb. In this paper we used the HSV color space that is developed to provide an intuitive representation of color and to approximate the way in which humans perceive and manipulate color [16]. The hue (H) represents the dominant spectral component color in its pure form, as in green, red, or yellow. Adding white to the pure color changes the color: the less white, the more saturated the color is. The saturation (S) correspond to the less or more white saturated the color is. The value (V) corresponds to the brightness of color.

$$V = \max(R,\ G,\ B)$$

$$S = \begin{cases} \frac{V - \min(R,G,B)}{V} & \text{if } V \# 0 \\ 0 & \text{otherwise} \end{cases}$$

$$H = \begin{cases} 60(G - B)/(V - \min(R,G,B)) & \text{if } V = R \\ 120 + 60(B - R)/(V - \min(R,G,B)) & \text{if } V = G \\ 240 + 60(R - G)/(V - \min(R,G,B)) & \text{if } V = B \end{cases}$$

3.4 Histogram Local Binary Patterns

The operator of the local binary patterns (LBP) was proposed in the late 90s by Ojala et al. [5]. Extraction of LBP features is efficient and with the use of multi-scale filters; invariance to scaling and rotation can be achieve. The idea of this texture operator is to assign to each pixel a dependent code grayscale. The grey level of the center pixel (P_c) of coordinates (xc, yc) is compared with its neighbors (Pn) using the following Eq. (1). Figure 3 give an example:

$$\begin{aligned} \mathrm{LBP}\,(\mathrm{x_c},\,\mathrm{y_c}) &= \sum_{n=0}^{p} s(P_n - P_c) \\ s(P_n - P_c) &= 1 \text{ if } \mathrm{Pn} - \mathrm{P_c} \geq 0 \\ &= 0 \text{ if } \mathrm{Pn} - \mathrm{P_c} < 0 \end{aligned} \qquad ((1))$$

where p is the number of neighboring pixels. In general, we consider a neighborhood of 3 * 3 where p = 8 neighbors. So we get, as an image to grayscale, a matrix containing LBP values between 0 and 255 for each pixel. A histogram is calculated based on these values to form the LBP descriptor.

For our descriptor, we used the uniform LBP, which extracts the most fundamental structure from the LBP. A LBP descriptor is considered to be uniform if it has at most two 0–1 or 1–0 transitions. For example, the pattern 00001000 (2 transitions) and 10000000 (1 transition) are both considered to be uniform patterns since they contain at most two 0–1 and 1–0 transitions. The pattern 01010010 on the other hand is not considered a uniform pattern since it has six 0–1 or 1–0 transitions.

Based on this, we propose using those nine uniform patterns that have a U value of at most 2 (00000000, 00000001, 00000011, 00000111, 00001111, 00011111, 00111111, 01111111, and 11111111). These nine patterns correspond to 58 of the 256 original unrotated patterns that can occur in the 3 × 3 neighborhood. Remaining patterns are accumulated into a single bin, resulting in a 59-bin histogram.

Using only 58/256 of the pattern information may appear as a waste of information, but this approximation is supported by a very important observation. Namely, the chosen nine uniform patterns seem to contribute most of the spatial patterns present in deterministic micro-textures.

3.5 Our Method

In order to take advantage of the robustness of the descriptors described before, and to overcome their limitations, we introduce our method: at the beginning we extract the hog descriptor from the entire image because this descriptor give local feature of shape, then we segmented every image into segment. For our case 16 segments in

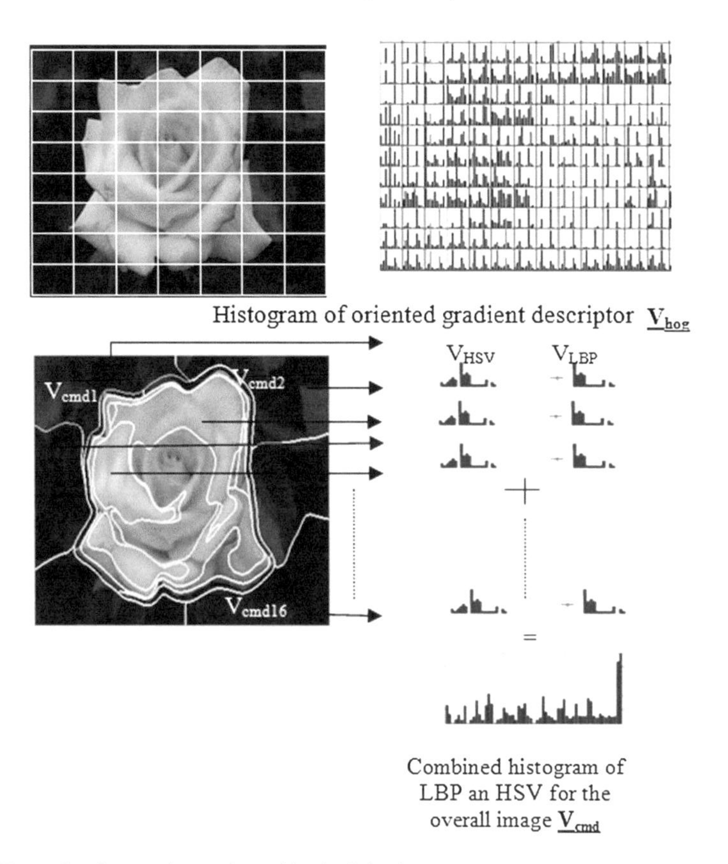

Fig. 4 Example of extraction and combined of the feature

order to minimize the density of information as shown in Fig. 2, and then we looping over each segment to extract and combine the HSV color and the LBP uniform histogram.. And for the calculation of similarity between vectors we used the Euclidean distance which proved very optimal for comparing vectors and histograms [15]. Figure 4 give an example:

Step 1: Extract the hog descriptor from the whole image: Vhog
Step 2: Applied the SLIC superpixel to the image to obtain 16 segments
Step 3: Loop over each superpixel and extract its contour.

- Compute bounding box of contour.
- Extract the rectangular ROI.
- Pass that into our descriptors to obtain our features
- Convert the segment to HSV space and extract histogram color: V_{HSV1}
- Convert the segment to grey and extract the LBP feature: V_{LBP1}
- Combined the two vectors:

$$V_{Cmd1} = V_{HSV1} + V_{LBP1}$$
$$\vdots$$
$$V_{Cmd16} = V_{HSV16} + V_{LBP16}$$

Step 4: concatenate the 16 combined descriptors of color and texture, to obtain visual local feature

$$V_{Cmd} = \{V_{Cmd1};\ V_{Cmd2};\ V_{Cmd3};\ \ldots..;\ V_{Cmd16}\}$$

Multiplied the vector by weighting factor and added the feature of shape:

$$V_{image} = V_{hog} + w1 * V_{Cmd}(w_{1} = 0.3)$$

4 Experimental Results

In our study, we used Corel image databases; of nature scenes classified according to several themes, the sample images are displayed in Fig. 5: The Simplicity dataset is a subset of COREL image dataset. It contains a total of 1000 images, which are equally divided into 10 different categories, the image are with the size of 256 * 384 or 384 * 256.

To evaluate our methods described above, we have set up an image search system that extracts the visual signatures of each image of the database as a vector of digital values and stores it in a data file. The signature of the query image will be compared later to those stored in the file according to the Euclidean distance, and return images with zero minimum distance to see the query image. To measure the quality of image search system content, parameters precision and recall are conventionally used [16]. We define Ai as set of all relevant image results for a given query and Bi represents all the images result returned by the system. The precision is defined as percentage of retrieved images belonging to the same category as the query image:

$$\mathrm{Pi} = \frac{Ai \cap Bi}{Bi}$$

Our system is designed to return 16 pictures following a query image; for each query we calculate the average retrieval precision (ARP):

$$\mathrm{ARP} = \frac{1}{N} \sum_{i=1}^{N} Pi$$

where N is the size of testing category in dataset.

In this experiment, our proposed method is compared with other image retrieval approaches reported in the literature [17–20] on the Corel-1000 dataset. To evaluate

Fig. 5 The simplicity dataset is a subset of COREL

Table 1 Comparison of different image retrieval approach on Corel-1000

Class	Yu [17]	Subrm [18]	Irtaza [19]	Elalami [20]	Our method
Africa	57.00	69.57	65.00	72.60	**77.77**
Beaches	58.00	54.25	60.00	59.30	**63.50**
Building	43.00	**63.95**	62.00	58.7	58.12
Bus	**93.00**	89.65	85.00	89.10	70.62
Dinosaur	98.00	98.7	93.00	99.30	**100**
Elephant	58.00	48.8	65.00	70.2	**74.30**
Flower	83.00	92.3	94.00	92.8	**98.32**
Horses	68.00	89.45	77.00	85.6	**100**
Mountain	46.00	4.30	**73.00**	56.20	53.75
Food	53.00	70.90	81.00	77.20	**82.95**
Total ARP	65.70	72.51	75.00	76.10	**77.93**

the performance of our method, we chose randomly ten images from every class (100 image in global) and very image is turned as query and then the precision rate is computed among all the query images under the number of retrieved image is 16. The average precision of each category using our method and the other approaches are shown in Table 1.

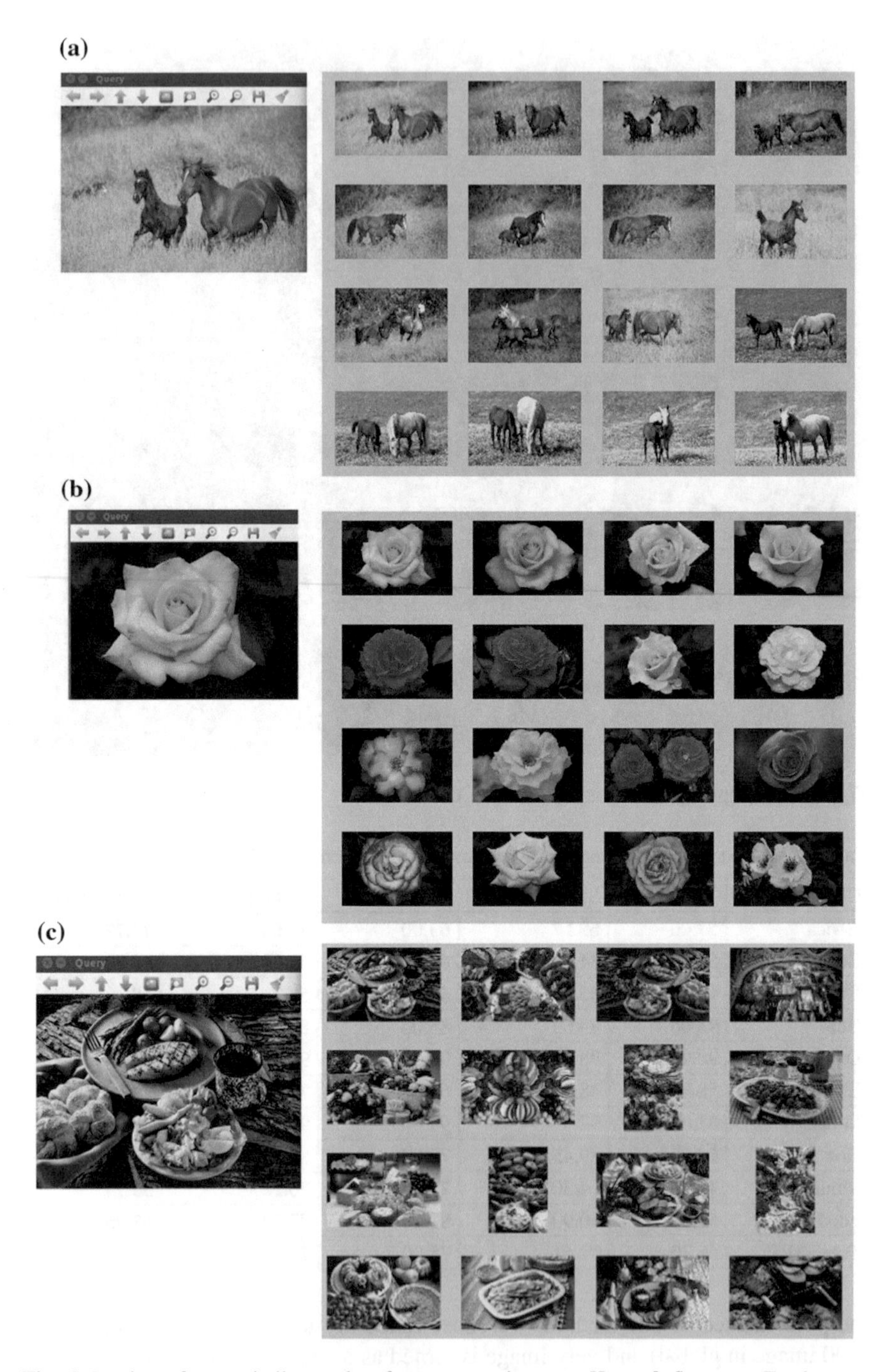

Fig. 6 Preview of some similar results of same query image. **a** Horse, **b** flowers, **c** Food

Table 1 shows the precision rate for each method, we observe better performance for our method in terms of precision, even though our method does not obtain the best performance in every class, but the majority (Africa, beach, Dinosaur, Flower, Horse, Food) have better result. The total average shows a good also better result 77.93 % compared to the other approach (76.10, 75.00….). Figure 6a–c show some results.

5 Conclusion

We have proposed a novel method for image retrieval using combination of color, texture and shape features. The hog descriptor is extract from the full image, and then the image is partitioned using superpixels segmentation, the LBP and HSV is drawn and combined for each segment, and regrouped in one vector. The vector hog is concatenate to the vector of the result vector of combined color and texture feature. The query image is treated with the same method, and the query vector is compared to the other in dataset, using the Euclidian distance. A combination of these descriptors provides a powerful feature set for image retrieval. The experiments using in the Corel dataset demonstrate the performance of this approach in comparison with the existing methods.

The effectiveness of a descriptor depends largely on the type of data and their heterogeneity, and the proposed combination in this work proved to be quite satisfactory and can give more performance on other types of base image. So it can be tested on other types of image-based to evaluate the performance of its results and bring him further improvement by combining different kind of descriptor.

References

1. X. Hu, G. Wang, H. Wu, H. Lu, Rotation-invariant texture retrieval based on complementary features, in *Proceedings of International Symposium on Computer, Consumer and Control*, 2014, pp. 311–314
2. M.J. Swain, D.H. Ballard, Color indexing. Int. J. Comput. Vision **7**(1), 11–32 (1991)
3. M.A. Stricker, M. Orengo, Similarity of color image, in *Proceedings of Storage an Retrieval for Image and Video Databases*, 1995, pp. 381–392
4. D.K. Park, Y.S. Jeon, C.S. Won, Efficient use of local edge histogram descriptor, in *Proceedings of ACM Workshops on Multimedia*, 2000, pp. 51–54
5. T. Ojala, M. Pietikainen, D. Harwood, A comparative study of texture measures with classification based on feature distribution. Pattern Recogn. **29**, 51–59 (1996)
6. D.G. Lowe, Distinctive image features from scale-invariant keypoints. Int. J. Comput. Vision **60**(2), 91–110 (2004)
7. Y. Ke, R. Sukthankar, PCA-SIFT: a more distinctive representation for local image descriptors, in *Proceedings of IEEE Conference on Computer Vision and Pattern Recognition*, vol. 2, 2004, pp. 506–513

8. H. Bay, A. Ess, T. Tuytelaars, L.V. Gool, SURF: speeded up robust features. Comput. Vis. Image Underst. **110**(3), 346–359 (2008)
9. L. Feng, J. Wu, S. Liu, H. Zhang, Global correlation descriptor: a novel image representation for image retrieval. Representation, 2015, pp. 104–114
10. N. Dalal, B. Triggs, Histograms of oriented gradients for human detection, in *IEEE Computer Society Conference on Computer Vision and Pattern Recognition. CVPR 2005*. IEEE, 2005, vol. 1, pp. 886–893
11. T. Malisiewicz, A.A. Efros, Improving spatial support for objects via multiple segmentations, 2007—repository.cmu.edu
12. X. Ren, J. Malik, Learning a classification model for segmentation, in *ICCV '03*, vol. 1, pp. 10–17, Nice 2003
13. G. Mori, X. Ren, A. Efros, J. Malik, Recovering human body configurations: combining segmentation and recognition, in *CVPR '04*, vol. 2, pp. 326–333, Washington, DC 2004
14. R. Achanta, A. Shaji, K. Smith, A. Lucchi, P.Fua, S. Susstrunk, SLIC Super-pixels; EPFL Technical Report 149300, 2010
15. H.Y. Lee, H.K. Lee, Y.H. Ha, Spatial color descriptor for image retrieval and video segmentation. IEEE Trans. Multim. **5**(3) (2003)
16. B.S. Manjunath, J.-R. Ohm, V.V. Vasudevan, A. Yamada, Color and texture descriptors. IEEE Trans. Circuits Syst. Video Technol. **11**(6) (2001)
17. J. Yu, Z. Qin, T. Wan, X. Zhang, Feature integration analysis of bag-of-features model for image retrieval. Neurocomputing **120**, 355–364 (2013)
18. M. Subrahmanyam, Q.M.J. Wu, R.P. Maheshwari, R. Balasubramanian, Modified color motif co-occurrence matrix for image indexing and retrieval. Comput. Electr. Eng. **39**, 762–774 (2013)
19. A. Irtaza, M.A. Jaffar, E. Aleisa, T.S. Choi, Embedding neural networks for semantic association in content based image retrieval. Multim. Tool Appl. **72**(2), 1911–1931 (2014)
20. M.E. ElAlami, A new matching strategy for content based image retrieval system. Appl. Soft Comput. **14**, 407–418 (2014)

DEPO: Detecting Events of Public Opinion in Microblog

Guozhong Dong, Wu Yang and Wei Wang

Abstract The rapid spread of microblog messages and sensitivity of unexpected events make microblog become the center of public opinion. Because of the large amount of microblog message stream and irregular language of microblog message, it is important to detect events of public opinion in microblog. In this paper, we propose DEPO, a system for Detecting Events of Public Opinion in microblog. In DEPO, abnormal messages detection algorithm is used to detect abnormal messages in the real-time microblog message stream. Combined with EPO (Events of Public Opinion) features, each abnormal message can be formalized as EPO features using microblog-oriented keywords extraction method. An online incremental clustering algorithm is proposed to cluster abnormal messages and detect EPO.

1 Introduction

Different from traditional news media, microblog allow users to broadcast short textual messages and express opinions using web-based or mobile-based platforms. Microblog provide the rapid communications of public opinion because of its immediacy, autonomy and interactivity. When emergency situation occurs, due to large number of people participating in conversation and discussions, some emergency situations which cause a surge of a large number of relevant microblog messages are named Events of Public Opinion (EPO) in this paper. Microblog messages related to events that have a significant increase or become popular during a certain time interval are called abnormal messages. In order to complete effective

G. Dong (✉) · W. Yang (✉) · W. Wang
Information Security Research Center, Harbin Engineering University, Harbin 150001, China
e-mail: dongguozhong@hrbeu.edu.cn

W. Yang
e-mail: yangwu@hrbeu.edu.cn

W. Wang
e-mail: w_wei@hrbeu.edu.cn

H. Lu and Y. Li (eds.), *Artificial Intelligence and Computer Vision*,
Studies in Computational Intelligence 672, DOI 10.1007/978-3-319-46245-5_6

management on public opinion of microblog after emergency situation occurred, it is necessary to detect and analyze EPO from microblog message stream by monitoring messages. EPO detection is broadly related to several areas: real-time system, social network analysis, parallel and distributed processing [1]. Unfortunately, events detection approach and system [2–5] have not been solved by the existing work on Chinese microblog. For example, trending topics list of microblog does not help much as it reports mostly those all-time popular topics, instead of EPO in our work. In this paper, we propose DEPO, an online EPO detection system. In DEPO, an online incremental clustering algorithm is used to cluster abnormal messages and detect EPO more accurately. Once burst events are detected, the system can summarize EPO and relevant abnormal messages.

2 System Overview

The system overview of DEPO, shown in Fig. 1, comprises three modules, namely Message Stream Distribution Module, Abnormal Messages Detection Module and EPO Detection Module.

Message stream distribution module is designed to handle massive real-time microblog messages. As real-time messages keep coming in, it enables DEPO to the distributed environment and constructs child message stream to abnormal messages detection module for further processing.

Abnormal messages detection module has several abnormal message monitor server, each abnormal message monitor server utilizes sliding time window model to divide and filter the message stream, only that the participation of message

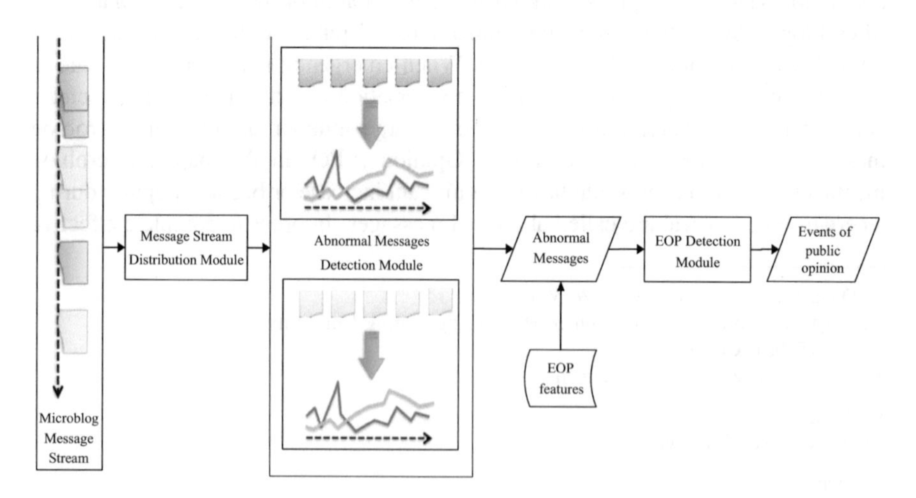

Fig. 1 The framework of DEPO

Table 1 The description of each tuple in microblog message

Description of each tuple	Formal representation
Message ID	*mid*
Original message ID	*root_mid*
User ID	*uid*
Comment number of original message	*com_num*
Retweet number of original message	*ret_num*
Post time of message	*post_time*
Post time of original message	*root_time*
Content of message	*content*

exceeds initial threshold is indexed in two-level hash table. Abnormal messages detection module computes each message influence series in hash table and determines whether it is an abnormal message in a given time window.

EPO detection module utilizes burst events detection algorithm combined with EPO features to cluster abnormal messages in each time window. The EPO features are labeled by 40 volunteers through labeling news section of Sina news.[1]

3 Methods

3.1 Sliding Time Window

Based on the transformation and storage of crawled microblog messages, microblog message m can be formalized as nine tuples. The description of each tuple is shown in Table 1.

A microblog message stream consist of microblog messages according to post time of messages which can be define as

$$M = [m_1, m_2, \ldots, m_i, \ldots, m_N] \tag{1}$$

If $i < j$ and $i, j \in \{1, 2, \ldots, N\}$, the post time of m_i is smaller than m_j.

The microblog message stream M can be divided into different time windows according to the post time of microblog message and time window size. Based on the concept of time window, the microblog message stream M can be formalized as

$$M = [W_1, \ldots, W_j, \ldots, W_L] \tag{2}$$

[1]http://news.sina.com.cn/.

where W_j represents the message set of j-th time window and $\sum_{j=1}^{L} |W_j| = M$. If W_L is current time window and K is the size of sliding window, sliding time window SW can be formalized as

$$SW = [W_{L-K+1}, \ldots, W_L] \tag{3}$$

3.2 Two-Level Hash Table

The two-level hash table is a kind of level hash tables. The brief structure of our proposed two-level hash table is shown in Fig. 2. It has two child hash tables T_0 and T_1. Each of them has corresponding hash function $hash_0(M)$ and $hash_1(M)$. These two functions are chosen from two global hash classes.

If $|T_0| = h_0$, $|T_1| = h_1$, $h_1 = h_0 \times r\ (0 < r < 1)$, $slot_{i,j}$ is the j-th slot position in child hash table T_j and $slot_{i,j} = T_{i,j}$, $i \in \{0, 1\}$, $j \in [0, h_i)$. The two child hash tables handle hash collision with separate chaining. $slot_{i,j}.list$ is the collision chain of $slot_{i,j}$. The length of collision chain in child hash table T_0 is limited and the

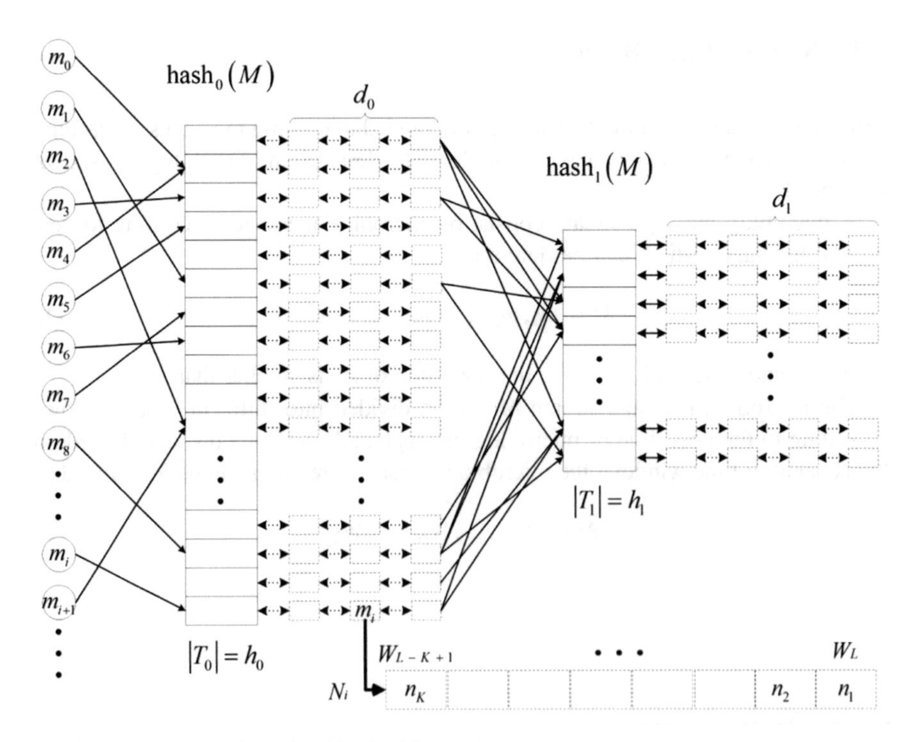

Fig. 2 The structure of two-level hash table

maximum length is set as d_0. The child hash table T_1 is a collision buffer of the child hash table T_0. Furthermore, each slot in T_0 has a *ofbuffer* which store the node information that the length of collision chain in child hash table T_0 is larger than d_0. Each message node in $slot_{i,j}.list$ can be formalized as

$$node = (m, N, pw) \tag{4}$$

where m is the microblog information stored in the node, pw is the number of past time windows, N is the message influence series in sliding time window. Message influence in current time window W_L can be computed as

$$n_1 = com\_num + ret\_num \tag{5}$$

where com_num is the number of comment in time window W_L, ret_num is the number of retweet in time window W_L.

Microblog message stream processing algorithm based on two-level hash table (Algorithm 1) can generate the message influence series of each message node in two-level hash table.

Algorithm 1. Microblog message stream processing algorithm

Input: M is microblog message stream, W_L is current time window, H is two-level hash table, $curr\_time$ is current time, $delay\_time$ is aging time

Output: Message influence series in sliding time window

```
(1)  for each mi ∈ M
(2)    if mi divided into WL
(3)      if mi is not in H
(4)        insert mi into H
(5)      else
(6)        if curr_time − mi.root_time > delay_time
(7)          update the node information of mi in H
(8)        else
(9)          delete mi from H
(10)   else
(11)     compute and store the influence series of mi
```

3.3 Abnormal Messages Detection

When current time window is full, a hash table copy signal is sent to abnormal messages detection thread. Abnormal messages detection algorithm based on two-level hash table (Algorithm 2) can detect abnormal messages in each time window. The algorithm uses the dynamic threshold adjustment strategy to set burst threshold, which can adjust burst threshold according to the message influence series.

Algorithm 2. Abnormal messages detection algorithm

Input: $N_i = (n_K, \cdots, n_2, n_1)$ is message influence series, HT is the threshold for hot messages, K is the size of sliding window
Output: M_h is the set of hot messages, M_b is the set of burst messages

```
(1)  if each node in two-level hash table
(2)    if node.pw < K
(3)      if node.m.com_num + node.m.ret_num > HT
(4)        add node to Mh and Mb
(5)    else
(6)      if node.m.com_num + node.m.ret_num > HT
(7)        add node to Mh
(8)      else
(9)        compute the Moving Average MAK of node.N
(10)        compute burst threshold BT
            BT = mean(MAK) + 2 * std(MAK)
(11)        if node.m.n1 > BT
(12)          add node to Mb
```

3.4 EPO Detection

EPO detection processes abnormal messages detected by all abnormal messages monitoring servers in each time window, which has two stages: abnormal messages pre-processing and abnormal messages clustering. In the stage of message pre-processing, user nickname and illegal characters in text content are first removed. Each abnormal message can be formalized as EPO features using microblog-oriented keywords extraction method. In the stage of abnormal messages clustering, we propose an abnormal messages incremental clustering algorithm to detect EPO in each time window. Combined with EPO features, the algorithm gives different weights to all kinds of features and computes similarity between abnormal messages. When the similarity between abnormal message and EPO is greater than merging threshold, they are grouped to a cluster.

4 Case Study

In this section, we show the case study of DEPO. We selected Sina microblog as observation platform in this section.

Abnormal messages detection. By using the abnormal messages detection method introduced in Sect. 3.3. Search interface for message retrieval are provided and users can retrieve abnormal messages according to the post time of messages and keywords. The sample abnormal messages search by a query on keywords can be seen in Fig. 3.

EPO detection. The interface of DEPO is shown in Fig. 4. Figure 4a shows events of public opinion detected by DEPO, including sensitive words, high frequency words, detecting time and alert level. Figure 4b shows the summary of EPO: the left of window shows the abnormal messages related to EPO, while the right of the window shows the statistical summary of EPO, such as geographical distribution, word cloud.

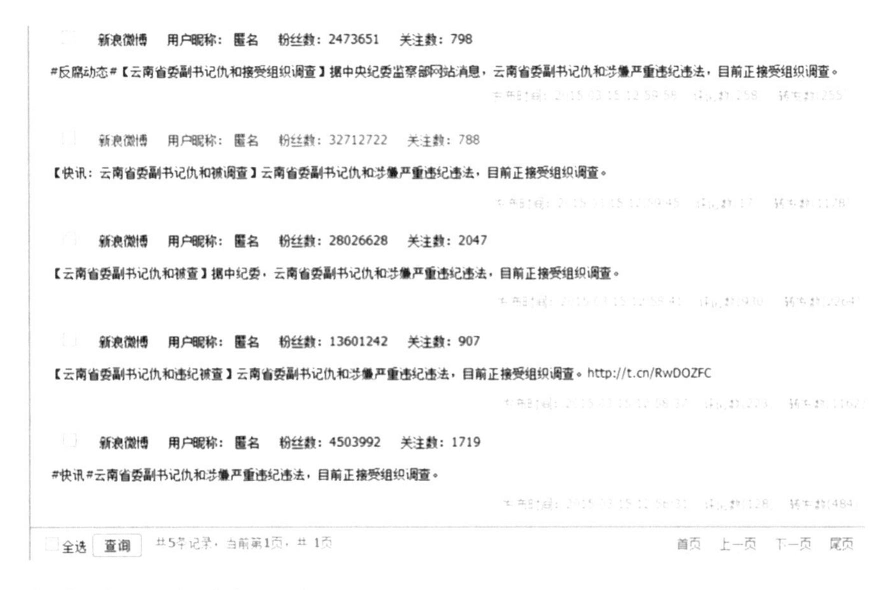

Fig. 3 The sample of abnormal messages

话题ID	敏感词	高频词	发现时间	更新时间	预警级别▼	详细内容	话题挖掘
5513	袭击,拆迁	衡阳,律师,衡阳市,回应,湖南,安徽	2015-04-21 10:44:25	2015-04-21 21:09:39	LV3	查看	点此进入
5631	肇事者,处罚	郑州,新郑,韩国,首尔	2015-04-21 20:47:29	2015-04-21 21:09:39	LV3	查看	点此进入
5479	矛盾,纠纷,立案,斗殴,执法,杀人,暴力	广东,澄海,中学,学生,湖北,红安,汕头,男生,视频,少年,调查,澄海区,斗殴,红安县	2015-04-20 16:07:00	2015-04-21 21:09:37	LV3	查看	点此进入
6023	曝光,龙卷风	济南,山东	2015-04-21 21:09:37	2015-04-21 21:09:37	LV3	查看	点此进入
5519	跳楼,身亡	河北,学生,青年报,衡水,教学楼,跳楼,学校,监狱,网友	2015-04-21 11:59:03	2015-04-21 21:05:40	LV3	查看	点此进入
6017	下药	男人,[illegible]	2015-04-21 21:05:28	2015-04-21 21:05:28	LV3	查看	点此进入
6019	曝光	余[illegible],上海,中国	2015-04-21 21:05:28	2015-04-21 21:05:28	LV3	查看	点此进入

(a) EPO list

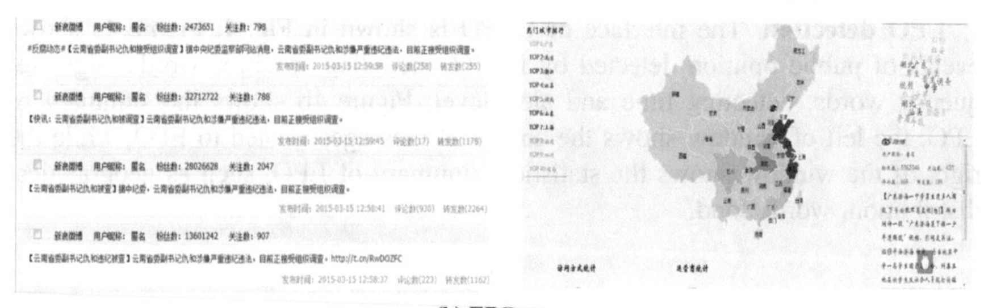

(b) EPO summary

Fig. 4 The interface of DEPO

Acknowledgments The authors gratefully acknowledge financial support from China Scholarship Council. This work was partially supported by the National High Technology Research and Development Program of China (no. 2012AA012802), the Fundamental Research Funds for the Central Universities (no. HEUCF100605) and the National Natural Science Foundation of China (no. 61170242, no. 61572459).

References

1. Y. Li, H. Lu, L. Zhang, J. Li, S. Serikawa, Real-time visualization system for deep-sea surveying. Math. Probl. Eng. (2014)
2. M.K. Agarwal, K. Ramamritham, M. Bhide, Real time discovery of dense clusters in highly dynamic graphs: identifying real world events in highly dynamic environments. Proc. VLDB Endow. **5**(10), 980–991(2012)
3. A. Cui, M. Zhang, Y. Liu, S. Ma, K. Zhang, Discover breaking events with popular hashtags in twitter, in *Proceedings of the 21st ACM International Conference on Information and Knowledge Management* (ACM, New York, 2012), pp. 1794–1798
4. C. Li, A. Sun, A. Datta, Twevent: segment-based event detection from tweets, in *Proceedings of the 21st ACM International Conference on Information and Knowledge Management* (ACM, Maui, 2012), pp. 155–164
5. R. Xie, F. Zhu, H. Ma, W. Xie, C. Lin, CLEar: a real-time online observatory for bursty and viral events. Proc. VLDB Endow. **7**(13), 1–4 (2014)

Hybrid Cuckoo Search Based Evolutionary Vector Quantization for Image Compression

Karri Chiranjeevi, Umaranjan Jena and P.M.K. Prasad

Abstract Vector quantization (VQ) is the technique of image compression that aims to find the closest codebook by training test images. Linde Buzo and Gray (LBG) algorithm is the simplest technique of VQ but doesn't guarantee optimum codebook. So, researchers are adapting the applications of optimization techniques for optimizing the codebook. Firefly and Cuckoo search (CS) generate a near global codebook, but undergoes problem when non-availability of brighter fireflies in search space and fixed tuning parameters for cuckoo search. Hence a Hybrid Cuckoo Search (HCS) algorithm is proposed that optimizes the LBG codebook with less convergence time by taking McCulloch's algorithm based levy flight distribution function and variant of searching parameters (mutation probability and step of the walk). McCulloch's algorithm helps the codebook in the direction of the global codebook. The variation in the parameters of HCS prevents the algorithm from being trapped in the local optimum. Performance of HCS was tested on four benchmark functions and compared with other metaheuristic algorithms. Practically, it is observed that the Hybrid Cuckoo Search algorithm has high peak signal to noise ratio and a fitness function compared to LBG, PSO-LBG, FA-LBG and CS-LBG. The convergence time of HCS-LBG is 1.115 times better to CS-LBG.

Keywords Vector quantization · Linde-Buzo-Gray (LBG) · Particle swarm optimization (PSO) · Firefly algorithm (FA) · Cuckoo search algorithm (CS) · Hybrid cuckoo search algorithm (HCS)

K. Chiranjeevi (✉)
Department of Electronics and Tele-Communication Engineering,
Veer Surendra Sai University of Technology, Burla, Odisha, India
e-mail: chiru404@gmail.com

U. Jena
Department of Electronics and Tele-Communication Engineering,
Veer Surendra Sai University of Technology, Burla, Odisha, India
e-mail: urjena@rediffmail.com

P.M.K. Prasad
Department of Electronics and Communication Engineering,
GMR Institute of Technology, Rajam, India
e-mail: mkprasad.p@gmrit.org

H. Lu and Y. Li (eds.), *Artificial Intelligence and Computer Vision*,
Studies in Computational Intelligence 672, DOI 10.1007/978-3-319-46245-5_7

1 Introduction

Image compression is concerned with minimization of the number of information carrying units used to represent an image. Due to the advances in various aspects of digital electronics like image acquisition, data storage and display, many new applications of the digital imaging have emerged; on the other hand many of these applications are not widespread because of the large storage space requirement. As a result, the importance for image compression grew tremendously over the last decade. Image compression plays a significant role in multimedia applications such as mobile, internet browsing, fax and so on. Now a day's establishment of image compression techniques with excellent reconstructed image quality is the crucial and challenging task. The image compression is aimed to transmit the image with lesser bitrates. The steps to image compression: identification of redundancies in image, proper encoding techniques and transformation techniques. Quantization is a powerful and efficient tool for image compression and is a non-transformed lossy compression technique. Quantization is classified into two types: Scalar quantization and Vector quantization (VQ). The aim of vector quantization is to design an efficient codebook. A codebook contains a group of codewords to which input image vector is assigned based on the minimum Euclidean distance. The Linde Buzo Gray (LBG) algorithm [1] is the primary and most used vector quantization technique. It is simple, adaptable and flexible, but suffers with local optimal problem; also it doesn't guarantee the global best solution. It is based on the minimum Euclidean distance between the image vector and the corresponding codeword. Patane proposed an enhanced LBG algorithm that avoids the local optimal problem [2]. Quad tree decomposition and the projection vector quantization (PVQ) provide variable sized blocks, but the performance of Quad tree decomposition projection is better than vector quantization (PVQ) [3]. Canta proposed a compression of multispectral images by address-predictive vector quantization based on identification and separation of spectral dependence and spatial dependence [4]. A quad tree (QT) decomposition algorithm allows VQ with variable block size by observing homogeneity of local regions [5]. But Sasazaki observed that complexity of local regions of an image is more essential than the homogeneity. So a vector quantization of images is proposed with variable block size by quantifying the complex regions of the image using local fractal dimensions (LFDs) [6]. Tsolakis proposed a Fuzzy vector quantization for image compression based on competitive agglomeration and a novel codeword migration strategy [7]. Tsekouras proposed an improved batch fuzzy learning vector quantization for image compression [8]. Comaniciu proposed an Image coding using transform vector quantization with a training set synthesis by means of best-fit parameters between input vector and codebook [9]. Wang observed that image compression can perform with transformed vector quantization. In this image to be quantized is transformed with discrete wavelet Transform (DWT) [10]. Tree-structured vector quantization (TSVQ) does not guarantee the closest codeword, So Chang proposed a full search equivalent TSVQ that generates an efficient closest codeword [11].

Recently soft computing techniques have grown in the fields of engineering and technology problems. Rajpoot designed a codebook by grouping the wavelet coefficients with the help of ant colony system (ACS) optimization algorithm. They designed a codebook by arranging the wavelet coefficients in a bidirectional graph and identification of the edges of the graph. They show that quantization of zero-tree vectors using ACS outperforms LBG algorithm [12]. Tsaia proposed a fast ant colony optimization for codebook generation by observing the redundant calculations [13]. In addition, Particle swarm optimization (PSO) vector quantization outperforms LBG algorithm which is based on updating the global best (gbest) and local best (lbest) solution [14]. The Feng showed that Evolutionary fuzzy particle swarm optimization algorithm has better global and robust performances than LBG learning algorithms [15]. Quantum particle swarm algorithm (QPSO) was proposed by Wang to solve the 0-1 knapsack problem [16]. The QPSO performance is better than PSO; it computes the local point from the pbest and gbest for each particle and updates the position of the particle by choosing appropriate parameters u and z. Poggi proposed Tree-structured product-codebook vector quantization, which reduces encoding complexity even for large vectors by combining the tree-structured component codebooks and a low-complexity greedy procedure [17]. Hu proposed a fast codebook search algorithm based on triangular inequality estimation [18]. Horng applied honey bee mating optimization algorithm for Vector quantization. HBMO has high quality reconstructed image and better codebook with small distortion compared to PSO-LBG, QPSO-LBG and LBG algorithm [19]. Horng [20] applied a firefly algorithm (FA) to design a codebook for vector quantization. The firefly algorithm has become an increasingly important tool of swarm intelligence that has been applied in almost all areas of optimization, as well as engineering practice [20]. Firefly algorithm is encouraged by social activities of fireflies and the occurrence of bioluminescent communication. Fireflies with lighter intensity values move towards the brighter intensity fireflies and if there is no brighter firefly then it moves randomly. So Chiranjeevi did some modification to FA called modified FA which follows a specific strategy instead random [21]. Chiranjeevi developed a bat algorithm which minimizes the mean square error between the input image and compressed image by means of bat algorithm based VQ [22]. Object-based VQ was proposed by Abouali based on an iterative process of LBG algorithm and max min algorithm, and Multi-object applications [23]. In this proposed work a hybrid cuckoo search algorithm is developed which takes the advantage of McCulloch's algorithm based L'evy distribution function and variation of parameters like mutation probability (Pa) and step of the walk (X) of ordinary cuckoo search. In ordinary cuckoo search algorithm L'evy distribution function follows Mantegna's algorithm and the tuning parameters like mutation probability (Pa) and step of the walk (X) are fixed. Cuckoo search algorithm with fixed parameters takes much time for convergence of problems, whereas Mantegna's algorithm based L'evy distribution function takes 80 % of the convergence time for local optima and remaining 20 % of convergence time for a global optimal solution. To address these two problems a hybrid cuckoo search algorithm is proposed.

This paper is organized in five sections including the introduction. In Sect. 2 recent methods of Vector Quantization techniques are discussed along with their algorithms. The proposed method of HCS-LBG algorithm is presented along with the procedure in Sect. 3. The results and discussions are given in Sect. 4. Finally the conclusion is given in Sect. 5.

2 Recent Methods of Codebook Design for VQ

The Vector Quantization (VQ) is a one of the block coding technique for image compression. Codebook design is an important task in the design of VQ that minimizes the distortion between reconstructed image and original image with less computational time. Figure 1 shows the encoding and decoding process of vector quantization. The image (size $N \times N$) to be vector quantized is subdivided into $N_b(\frac{N}{n} \times \frac{N}{n})$ blocks with size $n \times n$ pixels. These sub divided image blocks or training vectors of size n × n pixels are represented with X_i ($i = 1, 2, 3, \ldots N_b$). The Codebook has a set of codewords, C_i (where $i = 1 \ldots N_c$) is the ith codeword. The total number of codewords in Codebook is N_c. Each subdivided image vector is approximated by the index of codewords, based on the minimum Euclidean distance between corresponding vector and codewords. The encoded results are called an index table. During the decoding procedure, the receiver uses the same codebook to translate the index back to its corresponding codeword for reconstructing the image. The distortion D between training vectors and the codebook is given as

$$D = \frac{1}{N_c} \sum_{j=1}^{N_c} \sum_{i=1}^{N_b} u_{ij} \cdot \left\| X_i - C_j \right\|^2 \tag{1}$$

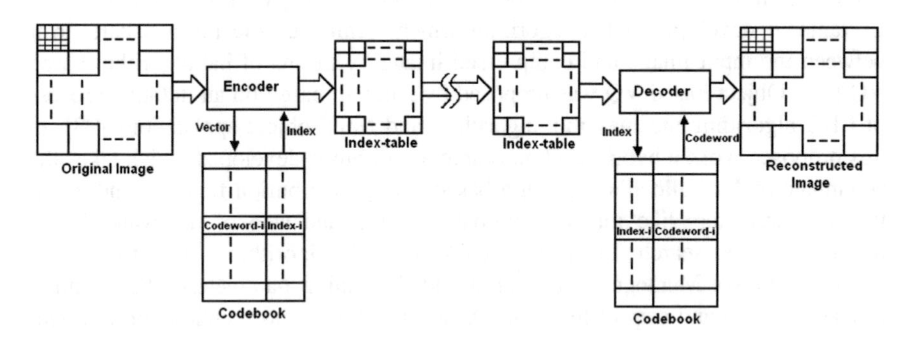

Fig. 1 Encoding and decoding process of vector quantization

Subject to the following constraints:

$$D = \sum_{j=1}^{N_c} u_{ij} = 1 \quad i = \{1, 2, \ldots, N_b\} \tag{2}$$

u_{ij} is one if X_i is in the jth cluster, otherwise zero. Two necessary conditions exist for an optimal vector quantizer. 1. The partition R_j, $j = 1, \ldots, N_c$ must satisfy

$$R_j \supset \{x \, \varepsilon \, X : d(x, C_j) < d(x, C_k), \forall k \neq j\} \tag{3}$$

where N_j is the total number of vectors belonging to R_j

2.1 Generalized LBG Vector Quantization Algorithm

The most commonly used methods in VQ are the Generalized Lloyd Algorithm (GLA) which is also called Linde-Buzo-Gray (LBG) algorithm. The algorithm is as follows:

Step 1 Begin with initial codebook C1 of size N. Let the iteration counter be m = 1 and the initial distortion D1 = ∞

Step 2 Using codebook Cm = {Yi}, partition the training set into cluster sets Ri using the nearest neighbor condition

Step 3 Once the mapping of all the input vectors to the initial code vectors is made, compute the centroids of the partition region found in Step 2. This gives an improved codebook Cm + 1

Step 4 Calculate the average distortion Dm + 1. If Dm—Dm + 1 < T then stops, otherwise m = m + 1 and repeat Step 2–4

The distortion becomes smaller after recursively executing the LBG algorithm. Actually, the LBG algorithm can guarantee that the distortion will not increase from iteration to the next iteration. However, it cannot guarantee the resulting codebook will become the optimum one and the initial condition will significantly influence the results. Therefore, in the LBG algorithm we should pay more attention to the choice of the initial codebook.

2.2 PSO—LBG Vector Quantization Algorithm

Kennedy proposed particle swarm optimization (PSO) based on social behavior of bird flocking or fish schooling [24]. There are two categories of PSO models: gbest and lbest models. The PSO gbest model was used by Zhao [25] to design a codebook for vector quantization by initializing the result of a LBG algorithm as gbest particle. In PSO particles/codebooks alter their positions/values based on their previous experience and the best experience of the swarm to generate a best codebook. Here codebook is assumed as a particle. The PSO algorithm is as follows:

Step 1 Run the LBG algorithm and assign it as global best codebook (gbest)
Step 2 Initialize rest codebooks with random numbers and their corresponding velocities
Step 3 Find out fitness values by Eq. (5) for each codebook

$$\mathrm{Fitness}(\mathrm{C}) = \frac{1}{\mathrm{D}(\mathrm{C})} = \frac{\mathrm{N_v}}{\sum_{j=1}^{\mathrm{N_c}} \sum_{i=1}^{\mathrm{N_b}} \mathrm{u_{ij}} \left\| \mathrm{X_i} - \mathrm{C_j} \right\|^2} \tag{5}$$

Step 4 If the new fitness value is better than old fitness (pbset) then assign its corresponding new fitness as pbest
Step 5 Select the highest fitness value among all the codebooks if it is better than gbest, then replace gbest with highest fitness value
Step 6 Update the velocities by Eq. (6) and update each particle to a new position by Eq. (7) and return to Step 3

$$v_{ik}^{n+1} = v_{ik}^{n} + c_1 r_1^n (pbest_{ik}^n - X_{ik}^n) + c_2 r_2^n (gbest_k^n - X_{ik}^n) \tag{6}$$

$$X_{ik}^{n+1} = X_{ik}^{n} + v_{ik}^{n+1} \tag{7}$$

where k is the number of solutions, i is the position of the particle, c_1, c_2 are cognitive and social learning rates respectively. r_1 and r_2 are random numbers.

Step 7 Until a stopping criterion is satisfied (Maximum iteration) repeat Step 3–7.

2.3 FA-LBG Vector Quantization Algorithm

Yang [26] introduced firefly algorithm (FA), inspired by the flashing pattern and characteristics of fireflies [26]. The brightness of a firefly equate to objective function value. The lighter firefly (lower fitness value) moves towards brighter firefly (higher fitness value). Here codebooks are assumed as fireflies. The detailed FA algorithm is given below.

Step 1 Run the LBG algorithm once and assign it as brighter codebook
Step 2 Initialize α, β and γ parameters, and rest codebooks with random numbers
Step 3 Find out fitness values by Eq. (5) of each codebook
Step 4 Randomly select a codebook and record its fitness value. If there is a codebook with higher fitness value, then it moves towards the brighter codebook (highest fitness value) based on the Eqs. (8)–(10)

$$Euclidean\,distance(r_{ij}) = \|X_I - X_J\| = \sqrt{\sum_{k=1}^{N_c} \sum_{h=1}^{L} (X_{i,k}^h - X_{j,k}^h)^2} \quad (8)$$

Here X_i is randomly selected codebook, X_j is brighter codebook

$$\beta = \beta_0 e^{-\gamma_{i,j}} \quad (9)$$

$$X_{j,k}^h = (1-\beta)X_{i,k}^h + \beta X_{j,k}^h + u_{j,k}^h \quad (10)$$

where u_{ij} is random number between 0 and 1, k = 1, 2, … , Nc, h = 1, 2, … L.
Step 5 If no firefly fitness value is better than the selected firefly then it moves randomly is search space with the following equation

$$X_{i,k}^h = X_{i,k}^h + u_{j,k}^h \quad k = 1, 2 \ldots N_c, h = 1, 2 \ldots L \quad (11)$$

Step 6 Repeat Step 3–5 until one of the termination criteria is reached.

2.4 CS-LBG Vector Quantization Algorithm

Yang developed a nature-inspired optimization algorithm called Cuckoo Search at Cambridge University to find the globally optimal solution for engineering problems [27]. It is inspired by the behavior and breeding process of cuckoo birds. Cuckoo search is not only used for linear problems, but also for nonlinear problems. Cuckoo birds emit beautiful sounds and its reproduction approach inspires the researchers. Cuckoo birds put down their eggs in the nests of host birds, if host bird recognizes that eggs are not its own, it will throw the alien eggs or abandons the nest and builds a new nest at some other location. Non-parasitic cuckoos, like most other non-passerines, lay white eggs, but many of the parasitic species lay colored eggs to match those of their passerine hosts. In some cases female cuckoo can mimic the color and pattern of eggs of some selected host nets. This feature minimizes the probability of eggs thrown away from the nest and causes an increment in productivity of cuckoos further. The cuckoos breeding process is based on the current position of cuckoo and probability of better next position after a selected random walk with a number of chosen random step size. This random walk plays a major role in the exploration, exploitation, intensification and diversification of the breeding process [28]. In general this foraging of random walk and random step size follows a probability density function which shows the distribution function of random walk. There are so many distribution functions like Gaussian distribution, normal distribution, L'evy distribution [29]. In cuckoo search optimization, random walk follows L'evy flight and step follows L'evy distribution function as in Eq. (14). L'evy flight is a random walk whose step follows the L'evy distribution function. In huge search space L'evy flight random walk is better than Brownian walk because of its nonlinear sharp variation of parameters [27]. The selection of random direction follows the uniform distribution function and generation of random walk steps by Mantegna algorithm which gives positive or negative numbers. L'evy distribution function is given as

$$L(s, \gamma, \mu) = \begin{cases} \sqrt{\frac{\gamma}{2\pi}} exp\left[-\frac{\gamma}{2(s-\mu)}\right] \frac{1}{(s-\mu)^{3/2}} & 0 < \mu < s < \infty \\ 0 & otherwise \end{cases} \tag{12}$$

where $\mu > 0$ is a minimum step and γ is the scale parameter. If $s \to \infty$ then Eq. (12) becomes

$$L(s, \gamma, \mu) \approx \sqrt{\frac{\gamma}{2\pi}} \frac{1}{(s)^{3/2}} \tag{13}$$

In cuckoo search algorithm mostly generation of random walk step is based on Mantegna's algorithm. According Mantegna's algorithm the step size of random walk of cuckoo is given by Eq. (14)

$$\text{Random walk step} = \frac{\mu}{(\nu)^{1/\beta}} \tag{14}$$

where μ and ν are drawn from normal distribution or Gaussian distribution is given in Eq. (15) with $\beta = 2$

$$L(s) = \frac{1}{\pi} \int_0^\infty \cos(\tau s) e^{-\tau\alpha^\beta} d\tau \tag{15}$$

From above equation μ and ν is given as

$$\mu \approx N\left(0, \sigma_{\mu}^{2}\right) \quad \nu \approx N\left(0, \sigma_{\nu}^{2}\right) \tag{16}$$

where the normal distribution function is given in Eq. (17)

$$N\left(\mu, \sigma^{2}\right) = \frac{1}{\sigma\sqrt{2\pi}} \exp\left[-\frac{(x-\mu)^{2}}{2\sigma^{2}}\right] \quad -\infty < x < \infty \tag{17}$$

where

$$\sigma_{\mu} = \left\{\frac{\Gamma(1+\beta)\sin\left(\frac{\pi\beta}{2}\right)}{\Gamma\left[\frac{1+\beta}{2}\right]\beta 2^{(\beta-1)/2}}\right\}^{\frac{1}{\beta}} \quad \text{and} \quad \sigma_{\nu} = 1 \tag{18}$$

where Γ is gamma function which is given in Eq. (19)

$$\Gamma(\beta) = \int_{0}^{\infty} e^{-t} t^{\beta-1} dt \tag{19}$$

Cuckoo search algorithm works with following three idealized rules: (1) each cuckoo lays one egg at a time, and dumps it in a randomly chosen nest; (2) The best nest with high quality of eggs (solutions) will carry over to the next generations; (3) The number of available host nests is fixed, and a host can discover an alien egg with a probability Pa $\in$ [0, 1]. In this case, the host bird can either throw the egg away or abandon the nest so as to build a completely new nest in a new location. The fitness function or fitness value is considered as objective function of problem. Here each Cuckoo nest is assumed as codebook. To optimize an efficient codebook, Cuckoo Search algorithm is applied which quantizes the input image efficiently. The detailed algorithm for vector quantization is as follows:

Step 1 (Initialization of parameters and solutions): Initialize number of host nests with each nest containing a single egg, a mutation probability (Pa) and a tolerance. Run the LBG algorithm and assign it as one of the nest/egg and rest nests are created randomly

Step 2 (selection of the current best solution): Calculate the fitness of all nests using Eq. (5) and select maximum fitness nest as current best nest nestbest

Step 3 (Generate new solutions with Mantegna's algorithm): New cuckoo nests (nestnew) are generated which are around the current best nest with a random walk (Lévy flight). This random walk follows Lévy distribution function which obeys Mantegna's algorithm. New nest is given as

$$\text{nestnew} = \text{nestold} + \alpha \otimes \text{L\'evy}(X) \tag{20}$$

where α is step size usually equal to 1 and Le'vy(X) is L'evy Distribution function obtained from the Eqs. (14) and (18). For the sake of simplicity Eq. (20) is modified as

$$\text{nestnew} = \text{nestold} + \text{step} \times (\text{nestbest} - \text{nest}) \tag{21}$$

where step is a random walk which follows Lévy distribution function Eq. (18)

Step 4 (discard worst nets and replace with new nests): If the generated random number (K) is greater than mutation probability (Pa) then replacing worse nests with new nests by keeping the best nest unchanged. New nests are

$$\text{nestnew} = \text{nestold} + (\text{K} \times \text{stepsize}) \tag{22}$$

generated by a random walk and random step size given as

$$\text{where Stepsize} = \text{r} \times (\text{nestrand} - \text{nestrand}) \tag{23}$$

Step 5 Rank the nests based on fitness function and select the best nest

Step 6 Repeat Step 2–6 until termination criteria.

3 Proposed Hybrid Cuckoo Search LBG Algorithm

PSO generates an efficient codebook, but undergoes instability in convergence when particle velocity is very high [30]. Firefly algorithm (FA) was developed to generate near global codebook, but it experiences a problem when no such significant brighter fireflies were available in the search space [31]. Cuckoo Search (CS) algorithm experiences a problem with fixed searching parameters and Mantegna's algorithm. So a hybrid cuckoo search (HCS) algorithm is proposed for global codebook. The HCS algorithm combines the advantages of L'evy distribution function obtained with McCulloch's algorithm and variation of parameters like mutation probability (Pa) and step of the walk (X) to overcome the disadvantage of ordinary Cuckoo Search which takes more iteration to construct a global solution. In normal Cuckoo search algorithm for optimizing a codebook the generation of random numbers follows Mantegna's algorithm based symmetric L'evy distribution function. However, recently it is found that L'evy distribution function obtained with McCulloch's algorithm is outperforming the Mantegna's algorithm and rejection algorithm [32]. So in our proposed method we applied McCulloch's algorithm L'evy distribution function. The draw back with the Mantegna's algorithm is that the convergence time of the algorithm slightly decreases with increment in X and 80 % of the convergence time is spent to extract the local optimal solution and remaining 20 % of the global optimal solution. Whereas McCulloch's algorithm L'evy distribution functions, convergence time is almost negligible for higher iterations and more efficient for large values of X and 20 % of the convergence time is spent to extract the local optimal solution and remaining 80 % of the global optimal solution.

In ordinary cuckoo search algorithm Pa and X are fixed. To get the best solution with fixed parameter cuckoo search algorithm takes more iteration. Valian proposed improved cuckoo search for global best solution by varying Pa and X [33]. Cuckoo search algorithm performance is poor with large value of X and small values of Pa. similarly, if X is small and Pa is large then the algorithm undergoes local optimal solution with less computational time. In order to overcome both the problems, concentration is paid on the way of adjusting Pa and X in the proposed algorithm so that a global solution is obtained. Initially the algorithm starts with high values of X and Pa so that algorithm can search for local solution and there after algorithm will decrease X and Pa values for global best solution. The variation of step size which follows McCulloch's algorithm is given in Eq. (24). It generates a step of random walk which depends on characteristic exponent α, skewness parameter β, scale c, and non-negligible parameter τ. We assume $\beta = 0$ and $\tau = 0$, m = 1and n = 16 (size of subdivided non overlapping block size) and N is current iteration count. Step of walk X is given as

$$step = X = c\left[\frac{\cos((1-\alpha)\varphi)}{w}\right]^{\frac{1}{\alpha}-1}\left[\frac{\sin(\alpha\varphi)}{\cos(\varphi)}\right]^{\frac{1}{\alpha}} \tag{24}$$

where

$$\varphi = \frac{(rand(m,n)) - 0.5) * \Pi}{N} \qquad w = \frac{-\log(rand(m,n))}{N}$$

Then finally new step of the walk is given as

$$X = \delta + (c * X) \tag{25}$$

The mutation probability (Pa) of the abandoned fraction of nests follows the following equation

$$Pa(n) = Pa\_\max - \frac{c}{N}(Pa\_\max - Pa\_\min) \tag{26}$$

3.1 Performance Evaluation of HCS Algorithm

In this section some comparisons between the proposed HCS, CS, FA and PSO using numerical benchmark test functions is demonstrated. The benchmark function chosen for validation of HCS are Ackley 1, Powell Singular 2, Alpine 1 and Brown functions [34]. The algorithm is validated with four variants of populations (P) and dimensions of search space (D) those are P = 30, D = 50; P = 30, D = 100; P = 50, D = 50 and P = 50, D = 100 described in Table 1. The variations in dimensions (D) help to check whether HCS gives better performance in both lower and higher dimensional search space. From Table 1 it is observed that HCS algorithm outperforms the other algorithms in lower and higher dimensional search space and also performance is better in lower and higher populations. All the experiments are performed 50 times. The simulation results of four benchmark functions i.e. f1, f2, f3 and f4 are shown in Fig. 2. This shows a graph between number of function evaluations and the objective function value. From Fig. 2. Minimum objective function value for the Ackley function with HCS is 0.5862, CS is 0.0409, FA is 1.2898 and PSO is 4.1826. With this example it can be concluded that HCS is better than other algorithms as its minimum value is near to the theoretical value '0'. Table 1 shows the results of four algorithms in the following aspects such as minimum, mean, and the standard deviation (std) of four benchmark functions. Here 'minimum' means the minimum value of objective function obtained in 50 independent runs. The word 'mean' means an average of 50 minimum objective function values. The word 'std.' implies the standard deviation of the best objective function values obtained from 50 independent runs. From Table 1, it can be observed that for all objective functions, the theoretical optima (minima) value is zeros. In this work, our objective is to find the global minimum. Hence, lower the 'minimum', 'mean' and 'std.', better is the algorithm. In this perspective, HCS performs well compared to CS, FA and PSO. The reason is that HCS takes the advantage of McCulloch's algorithm L'evy distribution function and advantage of variation in tuning parameters. Finally, the performances of all five algorithms are illustrated in Fig. 2. Figure 2 is drawn for objective function against 1000 functions evaluations. From Fig. 2a, one can compare performances of all four algorithms and four figures (Fig. 2a–d) are displayed for four benchmark functions separately. From Fig. 2, it is understood that the proposed algorithm HCS out-performs all other algorithms. In the following section, the proposed HCS is used for vector quantization for efficient codebook design. In this connection, a new algorithm called HCS-LBG is also proposed.

3.2 Proposed HCS-LBG Block Diagram

The block diagram of vector quantization using a Hybrid cuckoo search algorithm is shown in Fig. 3. An Image to be vector quantized is divided into immediate and non-overlapping blocks. These non-overlapping blocks are vector quantized with an LBG algorithm. The generated codebook of LBG algorithm is now trained with the hybrid cuckoo search algorithm. The trained codebook can satisfy the global convergence requirements and guarantee the global convergence properties. Furthermore, a hybrid cuckoo search is able to search local codebook and global codebook with the help of the variable control parameter. Assign any one of the codeword of trained codebook to non-overlapping blocks of the input image and its

Table 1 Minimum, mean and standard deviation of benchmark functions

Alg	f1(Akely)			f2(Powell)		
	Min	Mean	Std	Min	Mean	Std
HCS	0.0263	0.0624	0.0409	0.01	0.015	0.0344
	0.5862	1.0681	0.3288	0.069	0.131	0.0344
	0.8574	1.396	0.3684	11.45	15.46	1.5825
	0.1347	0.2408	0.1135	3.753	5.763	1.697
CS	0.0409	0.0791	0.0557	0.039	0.076	0.028
	0.7152	1.0824	0.21	0.457	0.777	0.2238
	1.5907	1.8161	0.1426	31.47	65.67	22.75
	0.2499	0.5748	0.3521	8.149	15.37	4.4225
FA	1.2898	1.4871	0.1476	234.7	270.6	28.732
	3.1861	4.2035	0.7063	214.4	257	27.897
	2.0402	2.2146	0.0821	10978	12513	110.31
	2.0098	2.1976	0.1093	11552	12726	87.95
PSO	4.1826	4.6867	0.3701	6833	15239	6269
	3.7495	5.0626	0.7455	12490	21689	5132
	4.7818	5.3846	0.4251	23182	44245	1122
	4.1809	4.8793	0.395	22586	34415	9217
	f3(Alpine)			f4(Brown)		
HCS	6.1008	6.8214	0.4681	0.392	0.688	0.2329
	4.2004	6.3615	1.0791	0.004	0.033	0.0237
	17.51	21.155	2.1854	18.13	34.87	18.827
	16.96	19.371	1.7618	4.57	16.32	16.321
CS	9.3565	10.906	1.1721	0.022	0.104	0.0677
	8.1832	10.368	1.2319	0.845	1.658	0.7099
	24.274	26.679	1.3042	34.54	137.2	236.6
	22.142	26.227	2.1745	11.3	44.24	57.159
FA	8.1832	10.368	1.2319	5.747	6.588	0.543
	3.2813	4.7363	0.9989	5.82	6.719	0.588
	11.024	12.76	1.0622	15.99	18.59	1.5788
	9.7005	11.941	1.1552	15.66	18.27	1.7025
PSO	10.47	14.192	2.9224	14.09	54.86	30.449
	7.519	11.1	2.0855	13.32	78.93	94.043
	23.547	27.513	3.054	53.28	337	243.96
	22.744	27.079	3.7694	76.19	134.4	48.654

corresponding index number form index table. These index table numbers are transmitted over the channel and decoded with the help of the decoder index table at the receiver. Rearrange all the decoded codewords in sequence such that the decompressed image size is same to that of the input image.

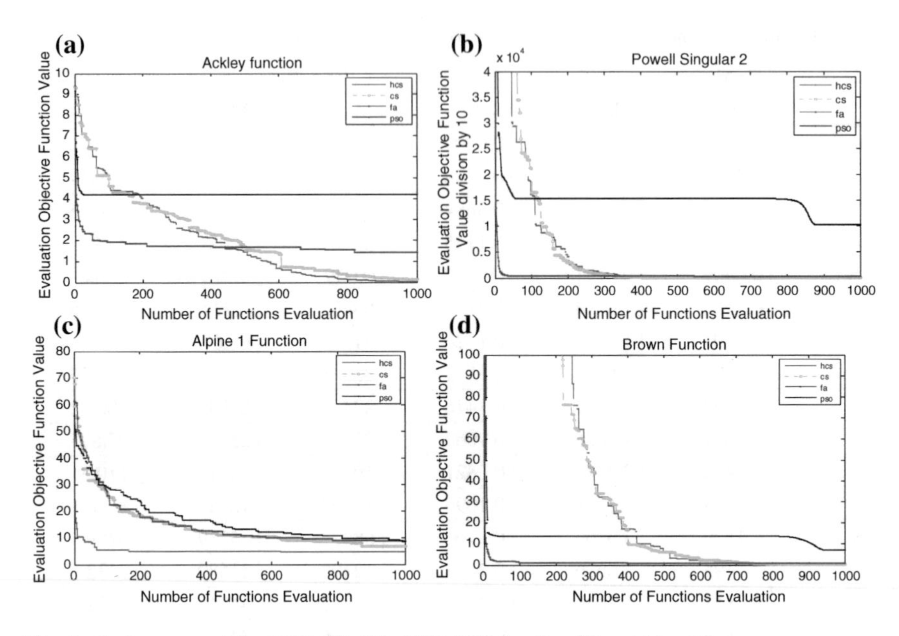

Fig. 2 Performance of the HCS, CS, FA AND PSO for P = 50 and D = 30

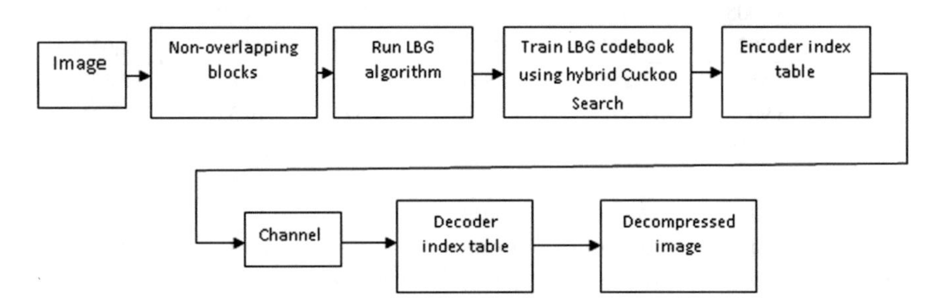

Fig. 3 Encoding and decoding process of vector quantization

4 Simulation Results and Discussion

We have chosen five different images: "LENA", "BABOON", "PEPPER", "BARB" and "GOLDHILL" for comparison of cuckoo search algorithm with other algorithms, as shown in Fig. 4a–e. All the images are grayscale images of size 512×512 pixels. Among all the images pepper is ".png" format and remaining images are ".jpg" format. The images are compressed with HCS-LBG, CS-LBG, FA-LBG, PSO-LBG and generalized Lloyd algorithm (LBG). As disused in Sect. 2, the image to be compressed is subdivided into non-overlapping images of size 4×4 pixels. Each subdivided image called as a block is treated as a training

Fig. 4 The five test images: **a** Lena, **b** Baboon, **c** Peppers, **d** Barb and **e** Goldhill

vector of 16 (4 × 4) dimensions. So there are 16384 input vectors to be encoded using a codebook as designed by one of the above algorithms.

The parameters used for comparison of proposed Hybrid cuckoo search algorithm with others are bitrate/bits per pixel, Peak Signal to Noise Ratio (PSNR) and Mean Square Error (MSE) as given in Eqs. (30)–(32) respectively. PSNR and fitness values are calculated for all the images with different codebook sizes of 8, 16, 32, 64, 128, 256, 512 and 1024. We use bpp (bit per pixel) to evaluate the data size of the compressed image for various codebook sizes of 8, 16, 32, 64, 128, 256, 512 and 1024. We then use the PSNR (peak signal-to-noise ratio) to evaluate the quality of the reconstructed image.

$$\text{bpp} = \frac{\log_2 N_c}{k} \tag{30}$$

where Nc is codebook size and k is the size of a block.

$$PSNR = 10 \times 10 \log\left(\frac{255^2}{MSE}\right) \text{dB} \tag{31}$$

where (MSE) which is given by the equation

$$MSE = \frac{1}{M \times N} \sum_{I}^{M} \sum_{J}^{N} \{f(\mathrm{I}, \mathrm{J}) - \bar{f}(\mathrm{I}, \mathrm{J})\}^2 \tag{32}$$

where M × N is the size of the image, I and J represents the coordinate values of pixel position of both the original and decompressed images. In our experiment we have taken N = M, i.e. a square image. $f(I, J)$ is the original image and $\bar{f}(I, J)$ is the reconstructed image.

The parameter values of the hybrid cuckoo search algorithm used for simulating the images are chosen based on the McCulloch's algorithm requirement and need a change in mutation probability and step of the random walk. The parameters used for simulating CS-LBG and HCS-LBG algorithms are sown in the Table 2. The parameters used for simulating PSO-LBG and FA-LBG are same as that referred in paper [20]. All the experiments are performed three times. To understand the

Table 2 The parameters of CS and HCS algorithm

Parameter	CS	HCS
Mutation probability (Pa)	0.55	0.55
Skewness parameter (β)	2	2
Delta (δ)	–	1
Characteristic exponent (α)	–	1.3
Scale (c)	–	0.1

performance of proposed method the graphs are drawn to show the variation of average peak signal to noise ratio (PSNR) against bitrate (BR).

Figures 5, 6, 7, 8 and 9 show the average peak signal to noise ratio of different tested images against bitrate. Experimentally it is shown that HCS algorithm improves the PSNR values by around 0.2 dB at low bit rates and 0.3 dB at higher

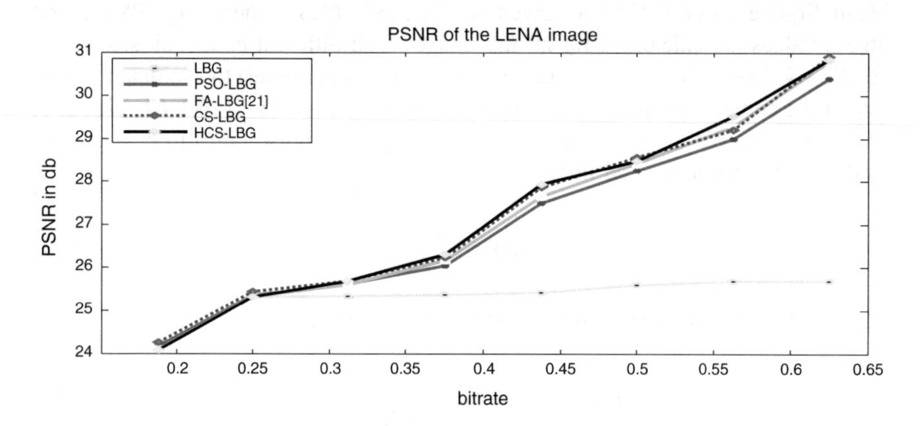

Fig. 5 PSNR of five VQ methods for LENA image

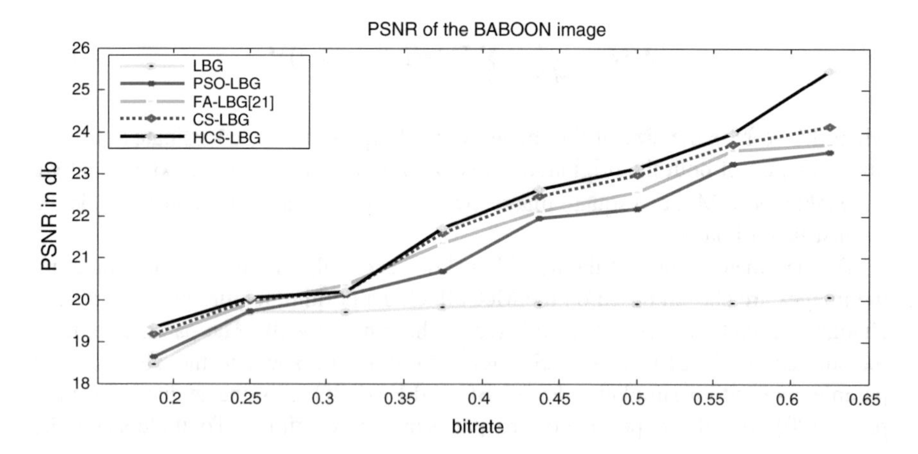

Fig. 6 PSNR of five VQ methods for Baboon image

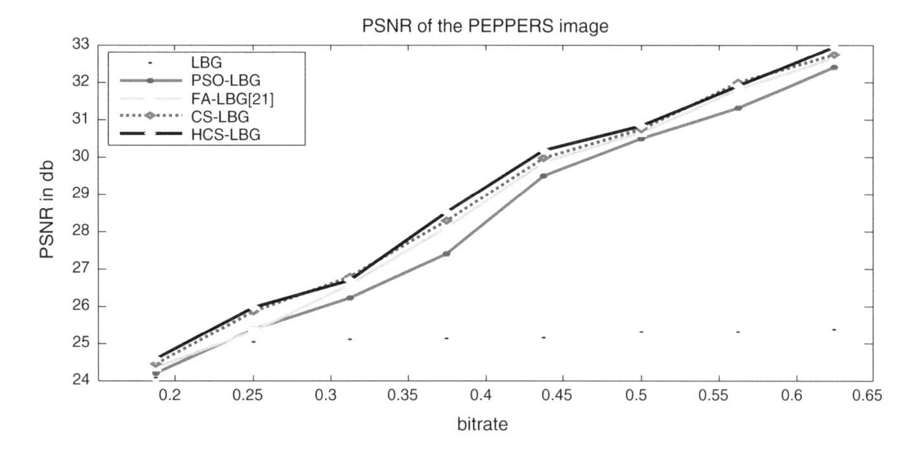

Fig. 7 PSNR of five VQ methods for Pepper image

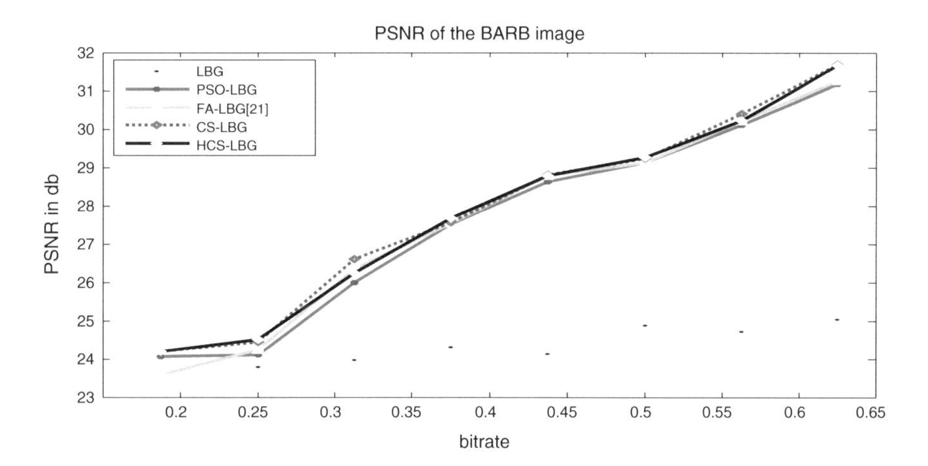

Fig. 8 PSNR of five VQ methods for Barb image

bit rates. HCS algorithm PSNR is better even for pepper image which has low spatial frequency components. Experimentally it is observed from the graphs that for different codebook sizes, HCS algorithm's PSNR value is better than LBG, PSO-LBG, FA-LBG and CS-LBG. These graphs reveal that for all algorithms PSNR value is better than the LBG algorithm.

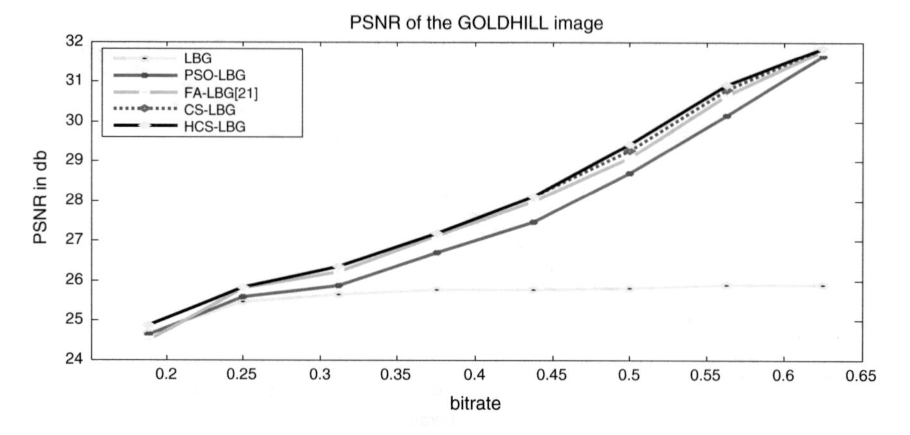

Fig. 9 PSNR of five VQ methods for Goldhill image

The empirical simulation was performed on Windows XP PC with an Intel(R) Core(TM) i5-2540 Machine with 2.60 GHz CPU, and 2.94 GB of RAM. Moreover, all the programs are written and compiled on MATLAB version 7.9.0 (R2009b). Tables 3, 4, 5, 6, 7, 8, 9 and 10 shows the average computation time or convergence time of different algorithms with different bitrates. Horng in his paper simulated the five algorithms in 'C++6.0' with windows XP operating

Table 3 Computation time with the bit rate = 0.1875. (Codebook size: 8)

Image	Average computation time (s)				
	LBG	PSO-LBG	FA-LBG	CS-LBG	HCS-LBG
Lena	3.37154	254.35792	877.12389	977.2009	977.20094
Pepper	3.41687	247.18151	660.14118	1019.698	1477.9607
Baboon	4.3135	322.99648	705.21004	1411.392	1032.3329
Goldhill	3.62287	247.21183	661.28155	1069.058	1069.0578
Barb	3.84336	257.52054	631.43537	1680.334	1265.7168
Average	3.71363	265.85365	707.0384	1231.536	1164.4538

Table 4 Computation time with the bit rate = 0.25. (Codebook size: 16)

Image	Average computation time (s)				
	LBG	PSO-LBG	FA-LBG	CS-LBG	HCS-LBG
Lena	3.40518	252.00159	507.18303	1678.34	1006.5839
Pepper	4.57563	250.41198	534.30463	1708.635	1185.5389
Baboon	4.53783	321.66919	943.36272	1455.553	1357.3564
Goldhill	4.90726	318.00435	574.98609	1261.561	1242.7106
Barb	4.39176	264.41467	737.33213	1337.834	1133.891
Average	4.36353	281.30036	659.43372	1488.385	1185.2162

Table 5 Computation time with the bit rate = 0.3125. (Codebook size: 32)

Image	Average computation time (s)				
	LBG	PSO-LBG	FA-LBG	CS-LBG	HCS-LBG
Lena	5.26223	306.53204	710.81085	1316.946	1316.9461
Pepper	6.2416	338.61886	594.78363	1090.372	990.75554
Baboon	5.17166	272.34614	723.56657	1579.7	1129.7974
Goldhill	4.48184	277.08706	755.60685	1525.981	1433.5292
Barb	6.67545	281.85315	877.69896	1346.776	1225.7281
Average	5.56655	295.28745	732.49337	1371.955	1219.3513

Table 6 Computation time with the bit rate = 0.375. (Codebook size: 64)

Image	Average computation time (s)				
	LBG	PSO-LBG	FA-LBG	CS-LBG	HCS-LBG
Lena	4.9586	311.44048	711.4527	1429.494	1429.4939
Pepper	5.76854	305.03441	633.62991	1531.719	1164.1182
Baboon	6.72856	314.85011	768.08809	1507.091	1330.0215
Goldhill	8.7956	382.04037	934.45318	1907.092	1151.6072
Barb	11.2127	308.21657	846.85393	1369.894	1221.5351
Average	7.4928	324.31639	778.89556	1549.058	1259.3552

Table 7 Computation time with the bit rate = 0.4375. (Codebook size: 128)

Image	Average computation time (s)				
	LBG	PSO-LBG	FA-LBG	CS-LBG	HCS-LBG
Lena	11.8885	522.28426	866.07975	1609.659	1536.959
Pepper	16.4215	507.58442	914.4514	1876.924	1418.0135
Baboon	19.6234	405.58237	920.11233	2195.839	1431.7771
Goldhill	15.2164	697.72016	1122.4419	1457.216	1247.4323
Barb	27.3468	521.94157	1145.6501	2044.023	1717.8457
Average	18.0993	531.02256	993.74711	1836.732	1502.3622

Table 8 Computation time with the bit rate = 0.5. (Codebook size: 256)

Image	Average computation time (s)				
	LBG	PSO-LBG	FA-LBG	CS-LBG	HCS-LBG
Lena	20.9133	875.94526	789.13847	2400.786	1597.3801
Pepper	18.1166	750.58442	972.20782	1707.318	1526.0773
Baboon	28.0281	594.62081	1032.4463	2006.82	1860.8022
Goldhill	29.7016	924.44431	827.91973	2932.647	2906.1385
Barb	27.6196	683.99975	830.79128	2555.872	2226.1864
Average	24.8758	765.91891	890.50072	2160.008	2183.9981

Table 9 Computation time with the bit rate = 0.5625. (Codebook size: 512)

Image	Average computation time (s)				
	LBG	PSO-LBG	FA-LBG	CS-LBG	HCS-LBG
Lena	56.317	1156.9086	1716.8064	2638.438	2618.1754
Pepper	82.0021	1950.2308	1357.8647	2862.149	2226.3954
Baboon	63.4333	2010.1972	2212.438	3544.229	3273.9501
Goldhill	77.8501	1291.1755	2126.3444	2776.047	2076.0466
Barb	115.211	1296.291	1386.5542	2868.87	2360.7334
Average	78.9627	1540.9606	1760.0015	2937.947	2511.0602

Table 10 Computation time with the bit rate = 0.625. (Codebook size: 1024)

Image	Average computation time (s)				
	LBG	PSO-LBG	FA-LBG	CS-LBG	HCS-LBG
Lena	145.726	3422.7993	4229.8094	8272.419	8145.1518
Pepper	140.347	2262.3369	2723.7381	4555.17	4191.6716
Baboon	156.577	3376.8045	3848.8403	5637.188	4523.774
Goldhill	181.405	2594.2074	2842.1809	4485.624	4679.7791
Barb	211.115	2847.8416	2221.6644	4511.959	4560.3902
Average	167.034	2900.7979	3173.2466	5492.472	5220.1534

systems, taking 100 numbers of codebooks/solutions and 50 numbers of iterations. The five algorithms are simulated in MATLAB with codebooks of 100 numbers and iterations of 50 numbers, so there is some dissimilarity of average computational time between proposed HCS-LBG and FA-LBG. From the observations of table, LBG algorithm computational time is very less compared to all other algorithms, but has lesser PSNR and bad reconstructed image quality. The average computation time of HCS algorithm is around 1.115 times faster than the CS-LBG. The normal fitness values of the five experimented images for five vector quantization algorithms are plotted in Figs. 10, 11, 12, 13 and 14. This investigation result confirms that the fitness of the five test images using the HCS-LBG algorithm is higher than the LBG, PSO-LBG, FA-LBG and CS-LBG. Figures 15, 16, 17, 18 and 19 shows the decompressed images of five different images for five vector quantization methods with a codebook size of 256 and block size of 16. It is observed that the decompressed/reconstructed image quality of the HCS-LBG algorithm is superior to the quality of reconstructed images of LBG, PSO-LBG, FA-LBG and CS-LBG.

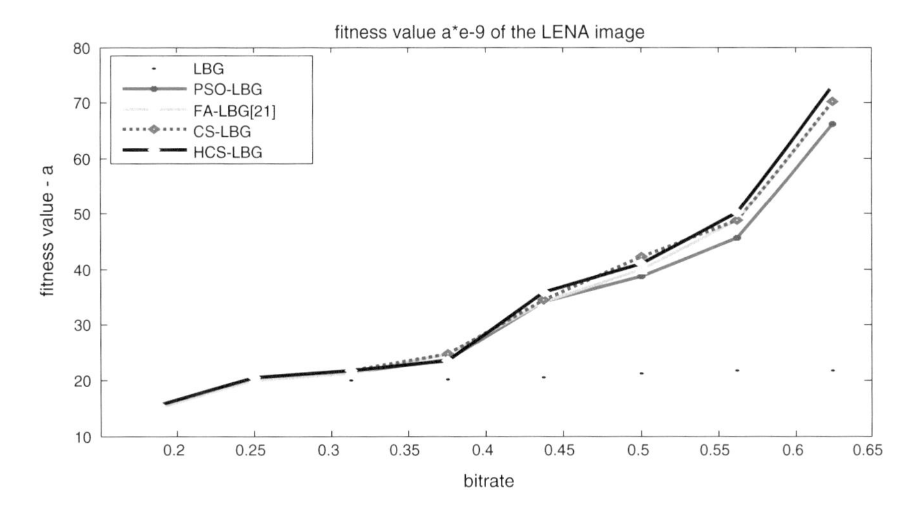

Fig. 10 Fitness values of five VQ methods for Lena

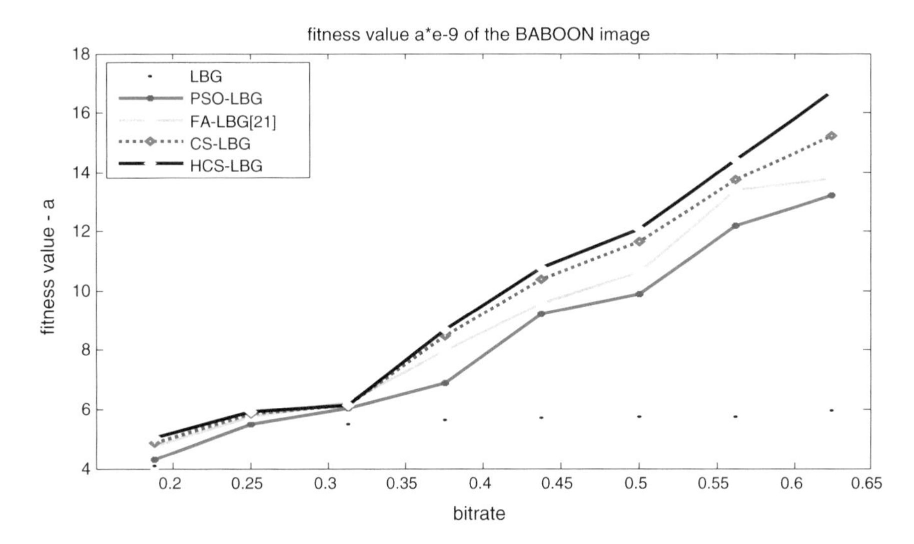

Fig. 11 Fitness values of five VQ methods for Baboon

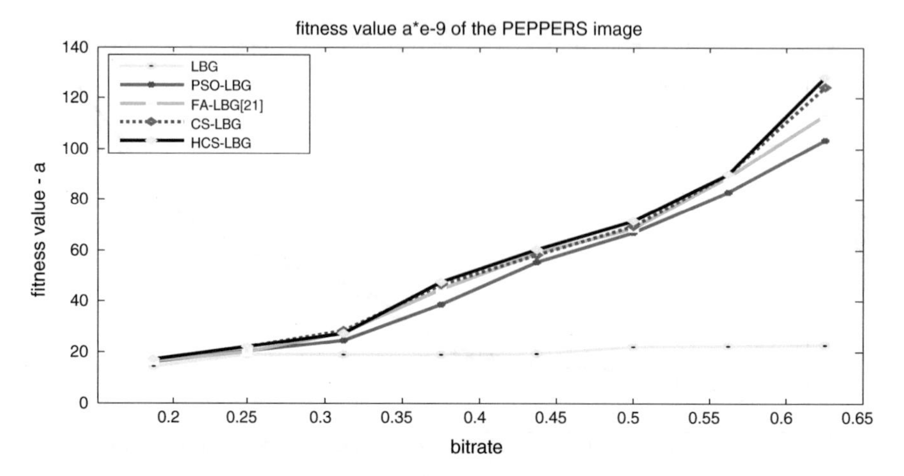

Fig. 12 Fitness values of five VQ methods for Peppers

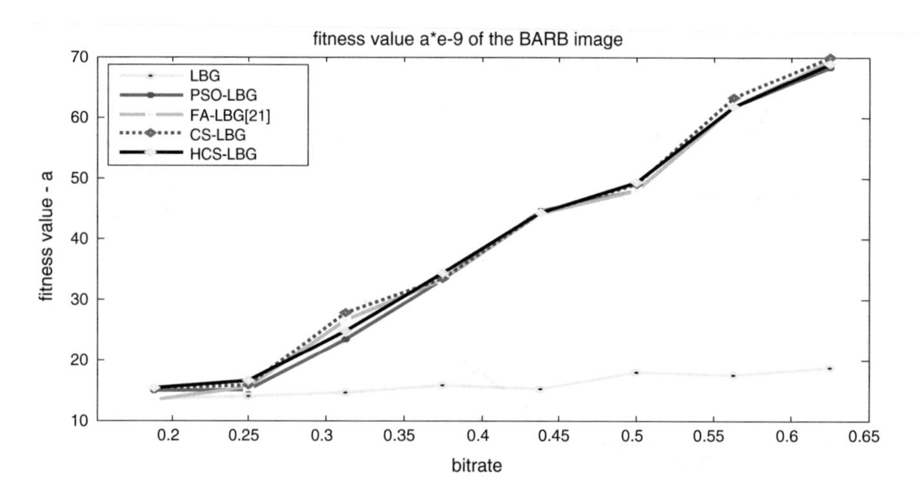

Fig. 13 Fitness values of five VQ methods for Barb

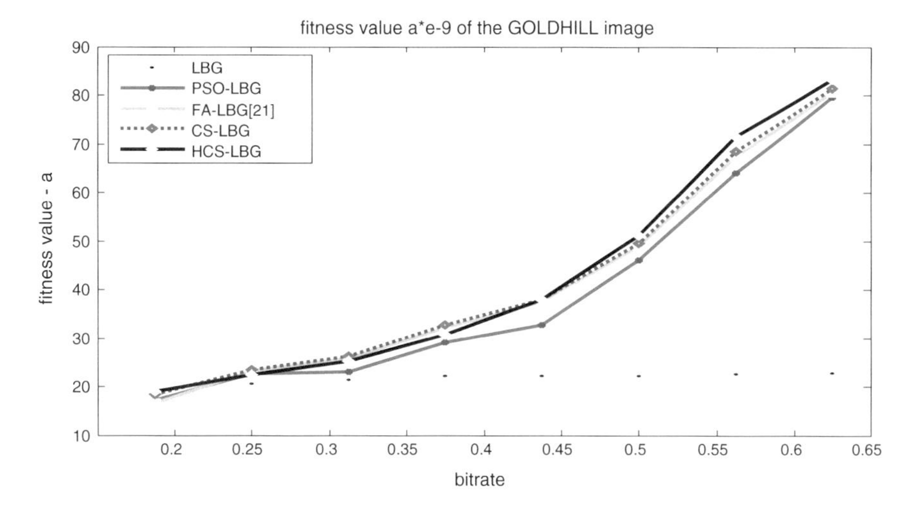

Fig. 14 Fitness values of five VQ methods for Goldhill

Fig. 15 Decompressed Barb with codebook size of 256 **a** LBG, **b** PSO, **c** FA, **d** CS, **e** HCS

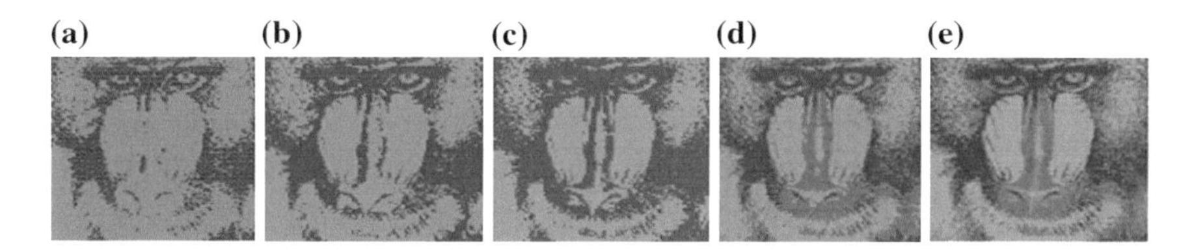

Fig. 16 Decompressed Baboon with codebook size of 256 **a** LBG, **b** PSO, **c** FA, **d** CS, **e** HCS

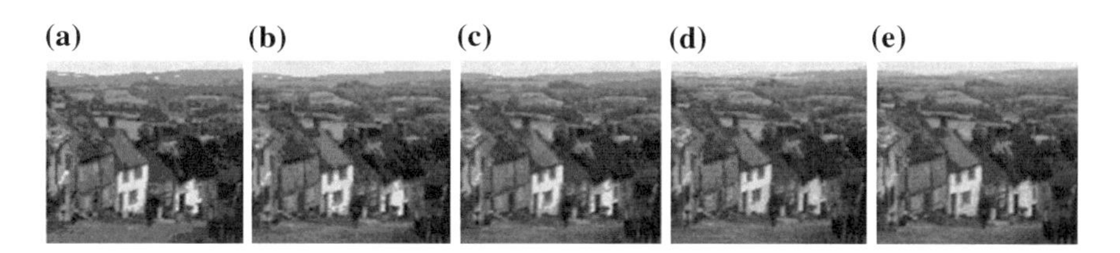

Fig. 17 Decompressed Goldhill with codebook size of 256 **a** LBG, **b** PSO, **c** FA, **d** CS, **e** HCS

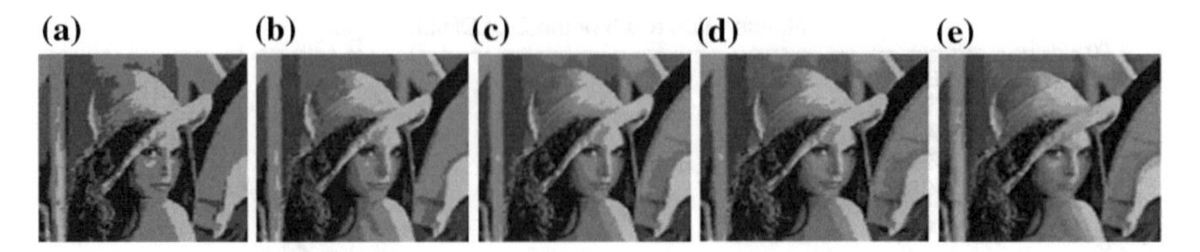

Fig. 18 Decompressed Lena with codebook size of 256 **a** LBG, **b** PSO, **c** FA, **d** CS, **e** HCS

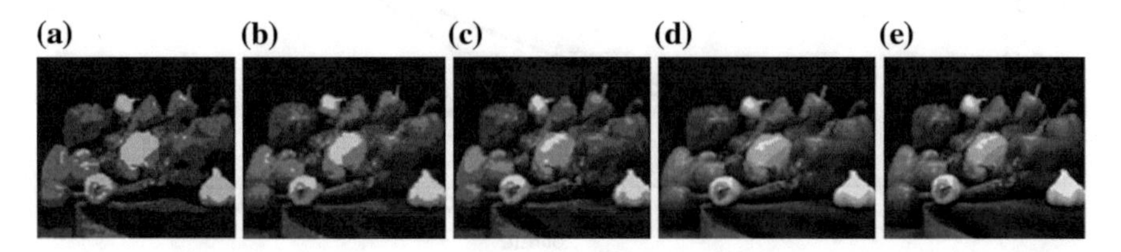

Fig. 19 Decompressed Pepper with codebook size of 256 **a** LBG, **b** PSO, **c** FA, **d** CS, **e** HCS

5 Conclusions

In this paper, a Hybrid cuckoo search algorithm based optimized vector quantization has been proposed for image compression. The Peak signal to noise ratio of vector quantization is maximized by employing HCS technique. The algorithm has been investigated by varying all possible parameters of HCS algorithms for efficient codebook design and efficient vector quantization of training vectors. Intensification and diversification of the algorithm are achieved by varying the mutation probability and step of walk. It is observed that the peak signal to noise ratio and quality of the reconstructed image obtained by HCS algorithm is superior to that obtained by LBG, PSO-LBG, FA- LBG and CS-LBG. From the simulation results it is observed that HCS-LBG has around 1.1115 times faster convergence rate than that of the CS-LBG. However, the HCS-LBG algorithm convergence speed not so good as compared to LBG, PSO-LBG and FA-LBG.

References

1. Y. Linde, A. Buzo, R.M. Gray, An algorithm for vector quantize design. IEEE Trans. Commun. **28**(1), 84–95 (1980)
2. G. Patane, M. Russo, The enhanced LBG algorithm. Neural Netw. **14**(9), 1219–1237 (2002)
3. K.H. Jung, C.W. Lee, Image compression using projection vector quantization with quad tree decomposition. Signal Process. Image Commun. **3**(5), 379–386 (1996)
4. G.R. Canta, G. Poggi, Compression of multispectral images by address-predictive vector quantization. Signal Process. Image Commun. **11**(2), 147–159 (1997)
5. Y.C. Hu, C.C. Chang, Quad tree-segmented image coding schemes using vector quantization and block truncation coding. Optim. Eng. **39**(2), 464–471 (2000)

6. K. Sasazaki, S. Saga, J. Maeda, Y. Suzuki, Vector quantization of images with variable block size. Appl. Soft Comput. **8**(1), 634–645 (2008)
7. D. Tsolakis, G.E. Tsekouras, J. Tsimikas, Fuzzy vector quantization for image compression based on competitive agglomeration and a novel codeword migration strategy. Eng. Appl. Artif. Intell. **25**(6), 1212–1225 (2012)
8. G.E. Tsekouras, M. Antonios, C. Anagnostopoulos, D. Gavalas, D. Economou, Improved batch fuzzy learning vector quantization for image compression. Inf. Sci. **178**(20), 3895–3907 (2008)
9. D. Comaniciu, R. Grisel, Image coding using transform vector quantization with training set synthesis. Signal Process. Image Video Coding **82**(11), 1649–1663 (2002)
10. X. Wang, J. Meng, A 2-D ECG compression algorithm based on wavelet transform and vector quantization. Digit. Signal Proc. **18**(2), 179–188 (2008)
11. C.C. Chang, Y.C. Li, J.B. Yeh, Fast codebook search algorithms based on tree-structured vector quantization. Pattern Recogn. Lett. **27**(10), 1077–1086 (2006)
12. A. Rajpoot, A. Hussain, K, Saleem, Q. Qureshi, A novel image coding algorithm using ant colony system vector quantization, in International Workshop on Systems, Signals and Image Processing (IWSSIP 2004), Poznan, Poland (2004)
13. C.W. Tsaia, S.P. Tsengb, C.S. Yangc, M.C. Chiangb, PREACO: a fast ant colony optimization for codebook generation. Appl. Soft Comput. **13**(6), 3008–3020 (2013)
14. Q. Chen, J.G. Yang, J. Gou, Image compression method by using improved PSO vector quantization, in Advances in Natural Computation, First International Conference on Neural Computation (ICNC 2005), Lecture Notes on Computer Science, vol. 3612, pp. 490–495 (2005)
15. H.M. Feng, C.Y. Chen, F. Ye, Evolutionary fuzzy particle swarm optimization vector quantization learning scheme in image compression. Expert Syst. Appl. **32**(1), 213–222 (2007)
16. Y. Wang, X.Y. Feng, Y.X. Huang, D.B. Pu, W.G. Zhou, Y.C. Liang, A novel quantum swarm evolutionary algorithm and its applications. Neurocomputing **70**(4), 633–640 (2007)
17. G. Poggi, A.R.P. Ragozini, "Tree-structured product-codebook vector quantization. Signal Process. Image Commun. **16**(20), 421–430 (2001)
18. Y.C. Hu, B.H. Su, C. Tsou Chiang, Fast VQ codebook search for gray scale image coding. Image Vis. Comput. **26**(5), 657–666 (2008)
19. M.H. Horng, T.W. Jiang, Image vector quantization algorithm via honey bee mating optimization. Expert Syst. Appl. **38**(3), 1382–1392 (2011)
20. M.H. Horng, Vector quantization using the firefly algorithm for image compression. Expert Syst. Appl. **39**(1), 1078–1091 (2012)
21. K. Chiranjeevi, J. Umaranjan, Modified firefly algorithm (MFA) based vector quantization for image compression, in Proceedings of the International Conference on Computational Intelligence in Data Mining (ICCIDM-2015) (Springer, 2015)
22. K. Chiranjeevi, J. Umaranjan, Fast vector quantization using a Bat algorithm for image compression. Eng. Sci. Technol. Int. J. **19**, 769–781 (2016)
23. A.H. Abouali, Object-based VQ for image compression. Ain Shams Eng. J. **6**(1), 211–216 (2015)
24. J. Kennedy, R.C. Eberhart, A new optimizer using particle swarm theory, in Proceedings of Sixth International Symposium on Micro Machine and Human Science, Nagoya, Japan, pp. 39–43 (1995)
25. Y. Zhao, A. Fang, K. Wang, H. Pang, Multilevel minimum cross entropy threshold selection based on quantum particle swarm optimization, in International Conference on Software Engineering Artificial Intelligence, Networking and Parallel/Distributed Computing, vol. 2, pp. 65–69 (2007)
26. X.S. Yang, *Nature-Inspired Metaheuristic Algorithms* (Luniver Press, 2008)
27. X.S. Yang, S. Deb, Cuckoo search via levy flights, in Proceedings of the World Congress on Nature and Biologically Inspired Computing, vol. 4, pp. 210–214 (2009)

28. C. Blum, A. Roli, Metaheuristics in combinatorial optimization: overview and conceptual comparison. ACM Comput. Surv. **35**(3), 268–308 (2003)
29. C. Brown, L.S. Liebovitch, R. Glendon, L´evy flights in Dobe Ju/'hoansi foraging patterns. Hum. Ecol. **35**(1), 129–138 (2007)
30. D.P. Rini, S.M. Shamsuddin, S.S. Yuhaniz, Particle swarm optimization: technique, system and challenges. Int. J. Comput. Appl. **14**(1), 0975–8887 (2011)
31. Q. Bai, Analysis of particle swarm optimization algorithm. Comput. Inf. Sci. **3**(1), 180–184 (2010)
32. H. Soneji, R.C. Sanghvi, Towards the improvement of Cuckoo search algorithm, in IEEE 2nd World Congress on Information and Communication Technologies (WICT-2012), pp. 878–883 (2012)
33. E. Valian, S. Tavakoli, S. Mohanna, A. Haghi, Improved cuckoo search for reliability optimization problems. Comput. Ind. Eng. **64**(1), 459–468 (2013)
34. T. Back, H.P. Schwefel, An overview of evolutionary algorithm for parameter optimization. Evol. Comput. **1**(1), 1–23 (1993)

Edge and Fuzzy Transform Based Image Compression Algorithm: edgeFuzzy

Deepak Gambhir and Navin Rajpal

Abstract Since edges contain symbolically important image information, their detection can be exploited for the development of an efficient image compression algorithm. This paper proposes an edge based image compression algorithm in fuzzy transform (F-transform) domain. Input image blocks are classified either as low intensity blocks, medium intensity blocks or a high intensity blocks depending on the edge image obtained using the Canny edge detection algorithm. Based on the intensity values, these blocks are compressed using F-transform. Huffman coding is then performed on compressed blocks to achieve reduced bit rate. Subjective and objective evaluations of the proposed algorithm have been made in comparisons with RFVQ, FTR, FEQ and JPEG. Results show that the proposed algorithm is an efficient image compression algorithm and also possesses low time complexity.

1 Introduction

Problem of storage and demand of exchanging images over mobiles and internet have developed large interest of researchers in image compression algorithms. Especially high quality images with high compression ratio i.e. low bit rate is gaining advantage in various applications such as interactive TV, video conferencing, medical imaging, remote sensing etc. The main aim of image compression algorithm is to reduce the amount of data required to represent a digital image without any significant loss of visual information. This can be achieved by removing as much redundant and/or irrelevant information as possible from the image without degrading its visual quality. A number of image compression methods exists in literature. Joint photographic experts group (JPEG), JPEG2000, fuzzy based, neural networks based,

D. Gambhir (✉) · N. Rajpal
School of Information and Communication Technology, Guru Gobind Singh Inderprastha University, Dwarka, New Delhi, India
e-mail: gambhir.deepak@gmail.com

N. Rajpal
e-mail: navin_rajpal@yahoo.com

H. Lu and Y. Li (eds.), *Artificial Intelligence and Computer Vision*,
Studies in Computational Intelligence 672, DOI 10.1007/978-3-319-46245-5_8

optimization techniques are the commonly used image compression techniques [1–7]. However, fuzzy logic's ability to provide smooth approximate descriptions have attracted researchers towards fuzzy based image compression methods. Fuzzy transform, motivated from fuzzy logic and system modeling, introduced by Perfilieva, possesses an important property of preserving monotonicity [8] that can be utilized significantly to improve the quality of compressed image. F-transform transforms an original function into a finite number of real numbers (called F-transform components) using fuzzy sets in such a way that universal convergence holds true.
Motivation: With the ever increasing demand of low bandwidth applications in accessing internet, images are generally exchanged at low bit rates. JPEG based on DCT is the most popularly used image compression standard. But at low bit rates, JPEG produces compressed images that often suffer from significant degradation and artifacts. Martino et al. [9] proposed an image compression method based on F-transform (FTR) that performed better than fuzzy relation equations (FEQ) based image compression and similar to JPEG for small compression rate. Later Petrosino et al. [10] proposed rough fuzzy set vector quantization (RFVQ) method of image compression that performed better than JPEG and FTR. Since F-transform has an advantage of producing a simple and unique representation of original function that makes computations faster and also has an advantage of preserving monotonicity that results in an improved quality of reconstructed compressed image, hence this paper proposes edge based image compression algorithm in F-transform domain named edgeFuzzy. The encoding of the proposed algorithm consists of following three steps:

1. **Edge detection using Canny algorithm**: In this step, each input image block is classified into either a low intensity (LI), a medium intensity (MI) or a high intensity (HI) block using canny edge detection algorithm.
2. **Intensity based compression using the fuzzy transform (F-transform)**: The blocks classified into LI, MI and HI blocks are compressed using the F-transform algorithm.
3. **Huffman coding**: The intensity based F-transform compressed image data is further encoded using Huffman coding technique to achieve low bit rate.

Contribution: It is well known that edges provide meaningful information present in an image. Thus, an image compression algorithm that exploits edge information is proposed. Input image blocks based on the number of edge pixels detected using canny edge detection algorithm are classified as either as LI, MI or HI blocks. Since LI blocks carry less information (because it contains less number of edge pixels) hence they can be compressed more as compared to MI and HI blocks using F-transform. Huffman coding is then performed on the compressed image, that further helps in reducing the achieved bit rate.The proposed algorithm produces a better visual quality of compressed image with well preserved edges. There is a significant improvement in the visual quality of compressed images obtained using the proposed algorithm as measured by PSNR over other state of art techniques as shown in Figs. 4, 5, 6, 7 and 8. The proposed algorithm also possess low time complexity as observed from Tables 1, 2, 3, 4, 5, 6, 7 and 8.

The rest of chapter is organized as follows: Literature review about the recent fuzzy based image compression algorithm is given in Sect. 2. F-transform based image compression is discussed in Sect. 3, and the proposed method is presented in Sect. 4. Results and discussions are provided in Sect. 5 and finally the conclusions are drawn in Sect. 6.

2 Literature Review

The process of image compression deals with the reduction of redundant and irrelevant information present in an image thereby reducing the storage space and time needed for its transmission over mobile and internet. Image compression techniques may either be lossless (reversible) or lossy (irreversible) [11]. In lossless compression techniques, compression is achieved by removing the information theoretic redundancies present in an image such that the compressed image is exactly identical to the original image without any loss of information. Run-length coding, entropy coding and dictionary coders are widely used methods for achieving lossless compression. Graphics interchange format (GIF), ZIP and JPEG-LS (based on predictive coding) are the standard lossless file formats. These techniques are widely used in medical imaging, computer aided design, video containing text etc. However, in lossy compression techniques, compression is achieved by permanently removing the psycho-visual redundancies contained in image such that the compressed image is not identical but only an approximation of the original image. Video conferencing and mobile applications are various applications using lossy image compression techniques. JPEG (based on DCT coding) is the most popularly used lossy image compression standard file format. Only lossy image compression techniques can lead to higher value of compression ratio.

Apart from providing semantically important image information, edges play an important role in image processing and computer vision. Edges contain meaningful data, hence their detection can be exploited for image compression. The main aim of edge detection algorithms is to significantly reduce the amount of data needed to represent an image, while simultaneously preserving the important structural properties of object boundaries. Du [12] proposed two algorithms for edge base image compression. First algorithm is based on transmission of SPIHT bit stream at encoder and detection of edge pixels in the reconstructed image whereas second algorithm is based on detection of edges at the encoder and their extraction followed by combination at the decoder. The clarity of edges is further improved by using edge enhancement algorithm. Desai et al. [13] proposed an edge based compression scheme by extracting edge and mean information for very low bit rate applications. Mertins [14] proposed an image compression method based on edge based wavelet transform. Edges are detected and coded as secondary information. Wavelet transform is performed in such a way that the previously detected edges are not filtered out and hence are successfully preserved in reconstructed image. Avramovic [15] presents a lossless image compression algorithm based on edge detection and local gradient.

The algorithm combines the important features of median edge detector and gradient adjusted predictor. In past few years, many edge detection image compression algorithm using fuzzy logic have been developed that are more robust and flexible compared to the classical approaches. Gambhir et al. [16] proposed adaptive quantization coding based image compression algorithm. The algorithm uses fuzzy edge detector for based on entropy optimization for detecting edge pixels and modified adaptive quantization coding for compression and decompression. Petrosino et al. [10] proposed a new image compression algorithm named as rough fuzzy vector quantization (RFVQ). This method is based on the characteristics of rough fuzzy sets. Encoding is performed by exploiting the quantization capabilities of fuzzy vectors and decoding uses specific properties of these sets to reconstruct the original image blocks. Amarunnishad et al. [17] proposed Yager Involutive Fuzzy Complement Edge Operator (YIFCEO) based block truncation coding algorithm for image compression. The method uses fuzzy LBB (Logical Bit Block) for encoding the input image along with statistical parameters like mean and the standard deviation. Gambhir et al. [18] proposed an image compression algorithm based on fuzzy edge classifier and fuzzy transform. Fuzzy classifier first classifies input image blocks into either a smooth or an edge block. These blocks are further compressed and decompressed using fuzzy transform (F-transform). The algorithm relies on automatic detection of edges in images to be compressed using fuzzy classifier. Here edge detection parameters once set for an image is assumed to be working with other images. Further, encoding a block to single mean value results in loss of information. Gambhir et al. [19] also proposed an algorithm named pairFuzzy that classifies blocks using competitive fuzzy edge detection algorithm and also reduces artifact using fuzzy switched median filter.

3 Fuzzy Transform Based Image Compression

Perfilieva proposed F-transform based image compression algorithm in [8, 20, 21] and compared its performance with the performance of JPEG and fuzzy relation equations (FEQ). Fuzzy transform converts a discrete function on the closed interval $[A, B]$ to a set of M finite real numbers called components of F-transform, using basis functions that forms the fuzzy partition of $[A, B]$. An inverse F-transform assigns a discrete function to these components, that approximates the original function up to a small quantity ϵ.

Fuzzy partition of the Universe:

Consider M $(M \geq 2)$ number of fixed nodes, $x_1 \leq x_2 \leq x_3 \cdots \leq x_M$, in a closed interval $[A, B]$ such that $x_1 = A$ and $x_M = B$. The fuzzy sets $[A_1, A_2, \ldots A_M]$ identified with their membership functions $[A_1(x), A_2(x), \ldots A_M(x)]$ defined on $[A, B]$ forms the fuzzy partition of the universe, if the following conditions hold true for $k = 1, 2, 3 \ldots M$.

1. $A_k(x)$ is a continuous function over the interval $[A, B]$.
2. $A_k(x_k) = 1$ and $A_k(x) = 0$ if $x \notin (x_{k-1}, x_{k+1})$.
3. $A_k : [A, B] \rightarrow [0, 1]$ and $\sum_{k=1}^{M} A_k(x) = 1$ for all x.
4. $A_k(x)$ increases monotonically on $[x_{k-1}, x_k]$ and decreases monotonically on $[x_k, x_{k+1}]$.

For equidistant set of M points, $[A_1, A_2, \ldots A_M]$ forms a uniform fuzzy partition if the following additional conditions are satisfied for all x and k=2, … $M-1(M\geq2)$:

a. $A_k(x_k - x) = A_k(x_k + x)$,
b. $A_{k+1}(x) = A_k(x - \delta)$ where $\delta = (x_M - x_1)/(M - 1)$.

3.1 Discrete Fuzzy Transform for Two Variables

Consider $(M + N)$ fixed nodes (where $M, N \geq 2$), $x_1, x_2, x_3, \ldots x_M$ and $y_1, y_2, y_3, \ldots y_N$ of a two dimensional function, $f(x, y)$ on closed interval $[A, B] \times [C, D]$ such that $x_1 = A$, $x_M = B$, $y_1 = C$ and $y_N = D$. Let $[A_1, A_2, A_3, \ldots A_M]$ be the fuzzy partition of $[A, B]$ identified with their membership functions $[A_1(x), A_2(x), \ldots A_M(x)]$ such that $A_i(x) > 0$ for $[i = 1, 2, 3, \ldots M]$ and $[B_1, B_2, B_3, \ldots B_N]$ be the fuzzy partition of $[C, D]$ identified with their membership functions $[B_1(y), B_2(y), \ldots B_N(y)]$ such that $B_j(y) > 0$ for $[j = 1, 2, 3 \ldots N]$. The discrete fuzzy-transform of the function $f(x, y)$ is then defined as:

$$F_{k,l} = \frac{\sum_{i=1}^{M} \sum_{j=1}^{N} f(x_i, y_j) A_k(x_i) B_l(y_j)}{\sum_{i=1}^{M} \sum_{j=1}^{N} A_k(x_i) B_l(y_j)} \tag{1}$$

for $k = 1, 2, 3, \ldots m$ and $l = 1, 2, 3, \ldots n$.

And the inverse discrete fuzzy transform of F with respect to $\{A_1, A_2, \ldots A_M\}$ and $\{B_1, B_2, \ldots B_N\}$ is defined as:

$$f_{FN}(i,j) = \sum_{k=1}^{m} \sum_{l=1}^{n} F_{k,l} A_k(x_i) B_l(y_j) \tag{2}$$

for $i = 1, 2, 3 \ldots M$ and $j = 1, 2, 3, \ldots N$.

Let $f(x, y)$ be an image with M rows and N columns. The discrete F-transform compresses this image $f(x, y)$ into F-components $F_{k,l}$ using the Eq. (1) for $k = 1, 2, \ldots m$ and $l = 1, 2, \ldots n$. The compressed image $f_{FN}(i,j)$ can be reconstructed using related inverse F-transform using Eq. (2) for $i = 1, 2, \ldots M$ and $j = 1, 2, \ldots N$. Perfilieva and Martino [9, 20] proposed a method of lossy image compression and its reconstruction based on discrete F-transform.

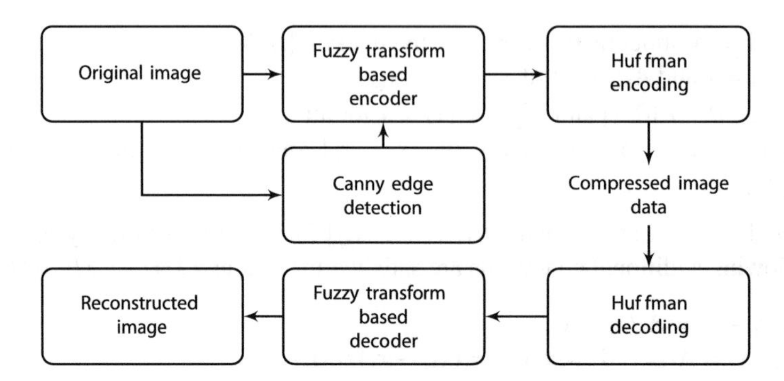

Fig. 1 Proposed method

4 Proposed Method

The proposed image compression method follows three steps:

1. Edge detection using Canny algorithm
2. Intensity based compression and decompression using F-transform
3. Huffman coding and decoding.

Figure 1 shows the block diagram for the proposed algorithm. The next subsections give the details of each step.

4.1 Edge Detection Using Canny Algorithm

Edge detection is a method of determining sharp discontinuities contained in an image. These discontinuities are sudden changes in pixel intensity which characterize boundaries of objects in an image. Canny edge detection [22], proposed by John F. Canny in 1986, is one of the most popular method for detecting edges. The performance of Canny detector depends upon three parameters: the width of the Gaussian filter used for smoothening the image and the two thresholds used for hysteresis threshold. Large width of the Gaussian function decreases its sensitivity to noise but at the cost of increased localization error and also some loss of detail information present in an image. The upper threshold should be set too high and the lower threshold should be set too low. Setting too low value of upper threshold increases the number of undesirable and spurious edge fragments in the final edge image and setting too high value of lower threshold results in break up of noisy edges. In MATLAB, lower threshold is taken to be 40 % of the upper threshold, if only the value of upper threshold is specified.

Fig. 2 Proposed fuzzy transform based encoder

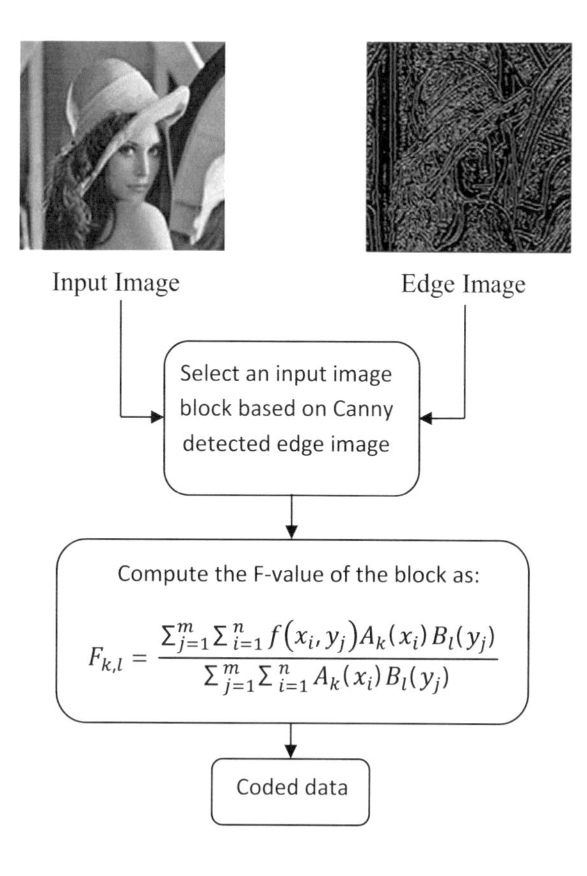

4.2 *Intensity Based Image Compression and Decompression Using F-Transform*

Monotonicity of a function is an important property preserved by F-transform that helps in improving the quality of compressed (reconstructed) image. Input image is first divided into blocks of size $n \times n$. Based on the edge image obtained from Canny edge detection algorithm, the input image blocks are classified into LI blocks, MI blocks and HI blocks. A block with small number of edge pixels (less than T1) is classified as LI block, with high number of edge pixels (greater than T2) as HI blocks and rest (with edge pixels between T1 and T2) as MI blocks. These blocks are further compressed using F-transform into different size blocks. Since LI blocks contain less information (as it contains less edge pixels) hence can be compressed more as compared to HI blocks. For example: a LI $n \times n$ block is compressed to 3×3 block, a MI $n \times n$ block is compressed to 5×5 block and a HI $n \times n$ block is compressed to 7×7 block. These compressed blocks are further encoded using lossless Huffman encoding to achieve lower bit rate. Figure 2 shows the proposed encoder.

4.3 *Introduction of Huffman Coding and Decoding*

The compressed image is further encoded using Huffman coding scheme to achieve more compression. Huffman code is a popular method used for lossless data compression introduced by Huffman [23], is optimum in sense that this method of encoding results in shortest average length. This coding technique is also fast, easy to implement and conceptually simple.
Summary: The proposed algorithm is summarized as: The proposed algorithm edgeFuzzy, creates an edge image using Canny edge detection algorithm and classifies input image blocks as either LI, MI or HI blocks based on this edge image. Based on the intensity, these blocks are then compressed into different size blocks using the F-transform. These compressed blocks are encoded using Huffman coding that further reduces the bit rate.The proposed algorithm can produce different bit rates depending on the number of edge pixels detected by the Canny algorithm. Too many edge pixels detected by the Canny algorithm will result in low compression and hence high bit rate. Thus the bit rate achieved by the algorithm is sensitive to the edge detection algorithm. Since Huffman coding is a lossless compression technique, therefore its utilization can further reduce the bit rate without any loss of visual information at the cost of minutely increased time complexity.

5 Results and Discussions

To reduce the storage space, bandwidth and time for uploading and downloading from internet and mobile, this paper proposes an edge based image compression algorithm in F-transform domain. The proposed algorithm exploits edge information extracted using Canny algorithm for compressing an image. Original images (row 1), Canny edge detected images with threshold T = 0.005 and width $\sigma = 1$ (row 2) and Canny edge detected images with threshold T = 0.2 and width $\sigma = 1$ is shown in Fig. 3. It is also observed that an increase in the value of threshold decreases the number of detected edge pixels. To evaluate the performance of the proposed algorithm, it has been tested on eight set of test images: Lenna, Bridge, House, Cameraman, Goldhill, peppers, Airplane and Lake of size 256×256 as well as on eight different set of test images: Tank, Straw, Aerial, Boat, Elaine, Lake, Pentagon and Wall of size 512×512. These set of images are downloaded from SIPI image database [24]. The process of compression is done on the PC with 4 GB RAM, Intel core i7 @ 2.50 GHZ with windows 8.1, 64 bit operating system using MATLAB 8.2, R2013b. The bit rate achieved using the proposed algorithm without lossless Huffman coding (i.e. bpp), using the proposed algorithm with lossless Huffman coding (i.e. bpp_H) and using the proposed algorithm with lossless arithmetic coding (i.e. bpp_A) in place of lossless Huffman coding is summarizes in Table 1 and Table 2 for images of size 256×256. These results are obtained by dividing input images of size 256×256 into blocks of size 16×16 and size 8×8 respectively and reducing

Fig. 3 *row 1*: Original images, Lenna, Bridge, House, Cameraman; Canny edge detected images at *row 2*: T = 0.005, *row 3*: T = 0.2 for the $\sigma = 0.5$

to 7×7 for HI block, 5×5 for MI block and 3×3 for LI block. The number of LI blocks, MI blocks and HI blocks for different images and different value of thresholds T1 and T2 is also shown in tables. It is also observed that small values of T1 and T2 increases the number of MI and HI blocks, this results in high bit-rate i.e. reduced compression. From these tables it is also observed that with the proposed algorithm different bit rate is achieved for different images at same values of T1 and T2. This is because compression using proposed method depends on the number of LI, MI and HI blocks and these number of blocks depends on the detected edge pixels. These number of detected edge pixels in turn depends on the parameters of Canny detection algorithm as well as on the nature of original image that is to be compressed. It is observed in Table 1 that the proposed algorithm achieved bit rate ranging from 0.032 bpp to 0.097 bpp approximately, while compressing original images of size 256×256 with the block size of 16×16. It is observed in Table 2 that the proposed algorithm achieved bit rate ranging from 0.118 bpp to 0.409 bpp approximately, while compressing original images of size 256×256 with the block size of 8×8. This increase in bit rate results in improvement of the visual image quality of the reconstructed image. Visual results of proposed algorithm for achieving compression of images of size 256×256 for blocks of size 16×16 (row 1–row 3) and for blocks of size 8×8 (row 4–row 6) is shown in Figs. 4 and 5 respectively.

Table 1 Proposed algorithm bit rate for T = 0.005 and $\sigma = 1$ for block size 16×16 (Image Size 256×256)

Images	T1	T2	LI	MI	HI	bpp	bpp_H	bpp_A
Lenna	0.40	0.50	256	000	000	0.035	0.032	0.028
	0.30	0.40	208	048	000	0.047	0.043	0.038
	0.20	0.35	070	181	005	0.082	0.075	0.069
Bridge	0.40	0.50	255	001	000	0.035	0.033	0.030
	0.30	0.40	144	111	001	0.063	0.058	0.053
	0.20	0.35	010	225	021	0.103	0.097	0.089
House	0.40	0.50	256	000	000	0.035	0.031	0.026
	0.30	0.40	183	073	000	0.053	0.047	0.042
	0.20	0.35	056	187	013	0.089	0.079	0.071
Cameraman	0.40	0.50	254	002	000	0.036	0.031	0.031
	0.30	0.40	136	118	002	0.065	0.056	0.054
	0.20	0.35	016	215	025	0.103	0.090	0.085
Goldhill	0.40	0.50	256	000	000	0.035	0.032	0.030
	0.30	0.40	169	087	000	0.056	0.050	0.046
	0.20	0.35	016	223	017	0.100	0.091	0.086
Peppers	0.40	0.50	256	000	000	0.035	0.033	0.029
	0.30	0.40	246	010	000	0.038	0.035	0.030
	0.20	0.35	098	157	001	0.074	0.070	0.062
Airplane	0.40	0.50	255	001	000	0.035	0.028	0.025
	0.30	0.40	195	060	001	0.050	0.039	0.035
	0.20	0.35	012	226	018	0.101	0.080	0.072
Lake	0.40	0.50	256	000	000	0.035	0.033	0.032
	0.30	0.40	221	035	000	0.044	0.040	0.039
	0.20	0.35	043	209	004	0.089	0.082	0.081

Alongwith the subjective evaluation, the proposed algorithm is also objectively evaluated using various quality measures such as: PSNR, RMSE and SAD. RMSE (root mean square error) [25] measures the square root of the cumulative squarer error between the original image and the compressed image. Mathematically

$$RMSE = \sqrt{\frac{\sum_{i=1}^{N} \sum_{j=1}^{M} [I(i,j) - C(i,j)]^2}{M \times N}} \tag{3}$$

where $M \times N$ is total number of pixels in an image. $I(i,j)$ and $C(i,j)$ are the intensity values of original and compressed images at location (i,j) respectively.

PSNR (peak signal to noise ratio) calculates the peak signal-to-noise ratio, in dB between two images and is defined as

Table 2 Proposed algorithm bit rate for T = 0.005 and $\sigma = 1$ for block size 8×8 (Image Size 256×256)

Images	T1	T2	LI	MI	HI	bpp	bpp_H	bpp_A
Lenna	0.40	0.50	1003	021	000	0.146	0.134	0.121
	0.30	0.40	813	190	021	0.200	0.185	0.170
	0.20	0.35	305	639	080	0.345	0.320	0.281
Bridge	0.40	0.50	970	053	001	0.154	0.146	0.129
	0.30	0.40	587	383	054	0.267	0.251	0.220
	0.20	0.35	116	718	190	0.432	0.409	0.369
House	0.40	0.50	978	046	000	0.152	0.136	0.121
	0.30	0.40	699	279	046	0.237	0.213	0.191
	0.20	0.35	246	635	143	0.383	0.345	0.317
Cameraman	0.40	0.50	959	065	000	0.156	0.137	0.120
	0.30	0.40	593	366	065	0.270	0.235	0.207
	0.20	0.35	118	685	221	0.443	0.387	0.341
Goldhill	0.40	0.50	983	040	001	0.151	0.139	0.122
	0.30	0.40	648	335	041	0.247	0.226	0.198
	0.20	0.35	147	735	142	0.407	0.374	0.352
Peppers	0.40	0.50	1016	008	000	0.143	0.135	0.118
	0.30	0.40	907	109	008	0.172	0.163	0.144
	0.20	0.35	402	595	027	0.302	0.287	0.253
Airplane	0.40	0.50	993	031	000	0.148	0.118	0.105
	0.30	0.40	672	321	031	0.238	0.184	0.165
	0.20	0.35	132	737	155	0.415	0.326	0.276
Lake	0.40	0.50	1010	014	000	0.144	0.134	0.121
	0.30	0.40	791	219	014	0.203	0.189	0.166
	0.20	0.35	256	681	087	0.360	0.336	0.298

$$PSNR = 20\ log_{10}\left(\frac{L}{RMSE}\right) \tag{4}$$

where L is the maximum possible value of intensity (for 8 bit image, $L = 255$).

SAD (sum of absolute difference) is used to measure the similarity between two images and is obtained using

$$SAD = \sum_{i=1}^{N}\sum_{j=1}^{M} |I(i,j) - C(i.j)| \tag{5}$$

Low value of RMSE, high value of PSNR and low value of SAD are generally desirable. Though these measures are most commonly used measures for objective analysis but these measures does not agree with human visual perception and hence SSIM and FSIM are also used for performance evaluation.

Fig. 4 Compressed images obtained using proposed algorithm (i) row 1 to row 3 for block size 16×16 and thresholds (0.4 and 0.5), (0.3 and 0.4) and (0.2 and 0.35) respectively and (ii) row 4 to row 6 for block size 8×8 and thresholds (0.4 and 0.5), (0.3 and 0.4) and (0.2 and 0.35) respectively for the test images of size 256×256

Fig. 5 Compressed images obtained using proposed algorithm (i) row 1 to row 3 for block size 16×16 and thresholds (0.4 and 0.5), (0.3 and 0.4) and (0.2 and 0.35) respectively and (ii) row 4 to row 6 for block size 8×8 and thresholds (0.4 and 0.5), (0.3 and 0.4) and (0.2 and 0.35) respectively for the test images of size 256×256

Table 3 Quality parameters obtained from the Proposed algorithm

Images	Code rate	PSNR (dB)	RMSE	SAD	MSSIM	FSIM
Lenna	0.032	24.40	15.36	728244	0.651	0.761
	0.043	24.78	14.70	695899	0.679	0.780
	0.075	25.91	12.91	576150	0.747	0.819
	0.134	27.87	10.31	425458	0.828	0.875
	0.185	28.82	9.24	378924	0.864	0.894
	0.320	30.26	7.82	281447	0.921	0.933
Bridge	0.033	21.46	21.55	1121689	0.390	0.614
	0.058	22.63	18.84	1053150	0.455	0.663
	0.097	23.57	16.90	885433	0.562	0.732
	0.146	24.63	14.96	790560	0.641	0.777
	0.251	25.71	13.33	676875	0.744	0.824
	0.409	26.96	11.44	516928	0.844	0890
House	0.031	22.49	19.14	931507	0.568	0.647
	0.047	24.82	14.64	880885	0.609	0.685
	0.079	25.11	14.15	756344	0.686	0.741
	0.136	25.92	12.89	611565	0.768	0.794
	0.213	27.43	10.84	532639	0.829	0.832
	0.345	29.71	8.33	403139	0.897	0.896
Cameraman	0.031	21.39	21.83	740882	0.667	0.701
	0.056	21.56	21.29	713780	0.690	0.719
	0.090	23.43	17.18	555940	0.767	0.779
	0.137	24.93	14.42	453574	0.830	0.857
	0.235	25.91	12.91	396270	0.870	0.857
	0.387	28.77	9.28	278423	0.929	0.914
Goldhill	0.032	23.10	17.84	835045	0.460	0.676
	0.050	23.52	16.91	787305	0.506	0.712
	0.091	24.89	14.53	662135	0.604	0.781
	0.139	25.84	13.03	588327	0.673	0.821
	0.226	26.76	11.70	521006	0.745	0.858
	0.374	28.47	9.61	416031	0.830	0.912
Peppers	0.033	22.05	20.13	821800	0.677	0.766
	0.035	22.09	20.04	817509	0.681	0.768
	0.070	23.44	17.16	668149	0.762	0.812
	0.135	26.25	12.41	450748	0.857	0.874
	0.163	26.68	11.81	427206	0.871	0.881
	0.287	28.65	9.42	323339	0.922	0.921

(continued)

Table 3 (continued)

Images	Code rate	PSNR (dB)	RMSE	SAD	MSSIM	FSIM
Airplane	0.028	25.44	13.63	353654	0.848	0.802
	0.039	25.46	13.59	349291	0.852	0.805
	0.080	27.27	11.03	271371	0.897	0.860
	0.118	28.23	9.88	241199	0.914	0.883
	0.184	28.48	9.60	227639	0.924	0.891
	0.326	30.15	7.93	173804	0.956	0.937
Lake	0.033	19.96	25.63	1108335	0.540	0.680
	0.040	20.12	25.21	1080084	0.555	0.689
	0.082	22.07	20.08	825933	0.685	0.764
	0.134	23.89	16.29	645148	0.772	0.827
	0.189	24.64	14.94	581306	0.815	0.851
	0.336	27.16	11.18	424595	0.894	0.910

SSIM (Structural similarity index measure) [26] measures the structural similarity between the two images and is calculated using:

$$SSIM = \frac{1}{W}\sum_{i=1}^{W}\left(\frac{2\mu_{i_i}\mu_{c_i} + (K_1L)^2}{\mu_{i_i}^2 + \mu_{c_i}^2 + (K_1L)^2}\right) \times \left(\frac{2\sigma_{i_ic_i} + (K_2L)^2}{\sigma_{i_i}^2 + \sigma_{c_i}^2 + (K_2L)^2}\right) \tag{6}$$

where μ_i, μ_c and σ_i, σ_c are mean intensities and standard deviations respectively, K_1 and K_2 are constants as, $0 < K_1, K_2 < 1$ and W is the number of local windows of the image. A large value of SSIM indicate the ability of algorithm to retain the original image.

The FSIM (feature similarity index measure) [27] measures the similarity between two images by computing locally the combination of the phase congruency (PC) [28] and gradient magnitude (GM) information using

$$FSIM = \frac{\sum_i \sum_j S(i,j) max\left\{PC_i(i,j), PC_c(i,j)\right\}}{\sum_i \sum_j max\left\{PC_i(i,j)PC_c(i,j)\right\}} \tag{7}$$

where

$$S(i,j) = \left(\frac{2PC_i(i,j)PC_c(i,j) + K_{PC}}{PC_i^2(i,j) + PC_c^2(i,j) + K_{PC}}\right) \times \left(\frac{2G_i(i,j)G_c(i,j) + K_{GM}}{G_i^2(i,j) + G_c^2(i,j) + K_{GM}}\right) \tag{8}$$

Table 4 Coding and decoding timing for test images (seconds)

Images	Code rate	Proposed coding time	Proposed decoding time	Code rate	FTR coding time	FTR decoding time	Code rate	JPEG coding time	JPEG decoding time
Lenna	0.032	3.52	1.58	0.035	4.10	14.82	0.034	4.13	3.31
	0.043	3.62	2.15	0.062	4.38	11.55	0.061	5.11	3.14
	0.134	3.79	2.40	0.140	5.46	14.93	0.130	5.87	3.61
	0.185	4.34	3.01	0.250	6.09	11.34	0.240	6.18	4.11
	0.320	4.95	3.73	0.391	6.21	11.72	0.439	6.24	4.28
Bridge	0.032	3.53	1.58	0.035	2.13	5.53	0.034	2.45	2.11
	0.043	3.61	2.13	0.062	1.19	4.19	0.058	6.78	2.36
	0.134	3.83	2.45	0.140	3.59	5.59	0.140	4.02	2.45
	0.185	4.29	3.11	0.250	2.32	4.32	0.244	4.65	2.68
	0.320	4.86	3.79	0.391	3.39	4.40	0.430	4.77	2.54
House	0.031	2.33	1.02	0.035	4.10	14.82	0.034	4.13	3.31
	0.047	3.35	1.54	0.062	4.38	11.55	0.062	5.11	3.14
	0.136	4.65	2.49	0.140	5.46	14.93	0.137	5.87	3.61
	0.213	3.53	1.68	0.250	6.09	11.34	0.240	6.18	4.11
	0.345	5.60	2.61	0.391	6.21	11.72	0.434	6.24	4.28
Cameraman	0.031	3.88	1.76	0.035	2.16	5.59	0.034	2.36	1.94
	0.056	5.66	2.74	0.062	1.20	4.25	0.061	5.54	2.09
	0.137	3.39	1.28	0.140	3.41	5.49	0.139	18.29	2.40
	0.255	4.08	1.86	0.250	2.11	4.31	0.249	3.07	2.68
	0.387	6.23	2.89	0.391	3.34	5.09	0.436	3.46	2.85

(continued)

Table 4 (continued)

Images	Code rate	Proposed coding time	Proposed decoding time	Code rate	FTR coding time	FTR decoding time	Code rate	JPEG coding time	JPEG decoding time
Goldhill	0.032	2.25	1.23	0.035	4.11	5.32	0.034	2.38	1.96
	0.050	3.63	1.95	0.062	4.68	4.44	0.061	5.56	3.12
	0.139	2.32	1.22	0.140	3.84	5.46	0.139	5.31	3.11
	0.226	3.76	1.93	0.250	5.24	4.51	0.249	3.16	2.32
	0.374	6.06	3.07	0.391	6.84	5.16	0.436	4.11	2.43
Peppers	0.033	2.31	1.23	0.035	2.16	5.59	0.034	4.33	1.94
	0.035	2.44	1.33	0.062	1.20	4.25	0.061	5.31	2.11
	0.135	2.12	1.15	0.140	3.41	5.49	0.139	7.03	3.10
	0.163	2.58	1.37	0.250	2.11	4.31	0.249	3.45	3.16
	0.287	4.55	2.28	0.391	3.34	5.09	0.436	4.15	2.50
Airplane	0.028	3.46	1.84	0.035	3.55	5.62	0.034	3.25	1.81
	0.039	4.77	2.62	0.062	4.96	5.25	0.061	4.15	2.11
	0.118	3.20	1.72	0.140	4.87	5.43	0.139	6.24	2.35
	0.184	5.21	2.77	0.250	6.69	4.28	0.249	3.25	2.64
	0.326	5.51	2.74	0.391	6.25	5.19	0.436	3.72	2.71
Lake	0.033	1.67	0.91	0.035	4.04	10.11	0.034	2.36	4.11
	0.040	2.35	1.21	0.062	3.03	9.82	0.061	5.54	4.15
	0.134	2.29	1.02	0.140	3.23	11.12	0.139	18.29	3.71
	0.189	2.62	1.41	0.250	3.45	14.18	0.249	3.07	4.59
	0.336	4.87	2.38	0.391	5.27	10.21	0.436	3.46	5.85

The corresponding values of these measures (PSNR, RMSE, SAD, SSIM and FSIM) for various images of size 256×256 at different code rates is shown in Table 3. It is also observed from the table, that low bit rate results degrade in quality of the reconstructed image. An idea about the time needed during coding and decoding images of size 256×256 using proposed algorithm, FTR and JPEG for achieving almost similar compression rates for various images when run on the same environment is given by Table 4. The proposed algorithm is much faster than its counterparts is also observed in Table 4.

The bit rate achieved using proposed algorithm (i.e. bpp_H) by dividing input images of size 512×512 into blocks of size 16×16 and size 8×8 is summarizes in Table 5 and Table 6 respectively. The results obtained using intensity based

Table 5 Proposed algorithm bit rate for T = 0.005 and $\sigma = 1$ for block size 16×16 (Image Size 512×512)

Images	T1	T2	LI	MI	HI	bpp	bpp_H	bpp_A
Elaine	0.40	0.50	1018	0006	0000	0.036	0.033	0.029
	0.30	0.40	0672	0346	0006	0.057	0.053	0.051
	0.20	0.35	0152	0755	0117	0.099	0.092	0.085
Boat	0.40	0.50	1022	0002	0000	0.035	0.031	0.026
	0.30	0.40	0728	0294	0002	0.053	0.046	0.042
	0.20	0.35	0096	0876	0054	0.097	0.085	0.081
Lake	0.40	0.50	1018	0006	0000	0.036	0.033	0.031
	0.30	0.40	0775	0243	0006	0.051	0.047	0.042
	0.20	0.35	0104	0840	0080	0.099	0.091	0.087
Straw	0.40	0.50	1022	0022	0000	0.036	0.030	0.028
	0.30	0.40	0207	0795	0022	0.087	0.073	0.069
	0.20	0.35	0000	0751	0273	0.123	0.104	0.095
Tank	0.40	0.50	1023	0001	0000	0.035	0.028	0.026
	0.30	0.40	0534	0489	0001	0.065	0.051	0.048
	0.20	0.35	0007	0969	0048	0.102	0.081	0.076
Aerial	0.40	0.50	1020	0004	0001	0.035	0.030	0.027
	0.30	0.40	0774	0246	0004	0.051	0.042	0.038
	0.20	0.35	0055	0926	0043	0.098	0.084	0.077
Wall	0.40	0.50	0990	0034	000	0.037	0.019	0.018
	0.30	0.40	0247	0743	0034	0.086	0.046	0.042
	0.20	0.35	0000	0672	0352	0.130	0.072	0.071
Pentagon	0.40	0.50	1024	0000	0000	0.035	0.028	0.027
	0.30	0.40	0773	0251	0000	0.050	0.040	0.036
	0.20	0.35	0062	0943	0019	0.096	0.077	0.074

Table 6 Proposed algorithm bit rate for T = 0.005 and $\sigma = 1$ for block size 8×8 (Image Size 512×512)

Images	T1	T2	LI	MI	HI	bpp	bpp_H	bpp_A
Elaine	0.40	0.50	3922	0174	0000	0.151	0.142	0.128
	0.30	0.40	2613	1309	0174	0.247	0.230	0.207
	0.20	0.35	0737	2717	0642	0.404	0.378	0.341
Lake	0.40	0.50	3930	0165	0001	0.151	0.140	0.126
	0.30	0.40	2797	1133	0166	0.235	0.216	0.195
	0.20	0.35	0826	2760	0600	0.395	0.366	0.330
Boat	0.40	0.50	3979	0117	0000	0.148	0.132	0.120
	0.30	0.40	2710	1269	0117	0.236	0.208	0.190
	0.20	0.35	0607	2951	0538	0.403	0.357	0.325
Straw	0.40	0.50	3646	0446	0004	0.168	0.145	0.131
	0.30	0.40	1394	2252	0450	0.347	0.304	0.274
	0.20	0.35	0066	2588	1442	0.519	0.458	0.413
Tank	0.40	0.50	3955	0141	000	0.149	0.120	0.117
	0.30	0.40	2247	1708	0141	0.266	0.214	0.197
	0.20	0.35	0189	3200	0707	0.444	0.359	0.340
Aerial	0.40	0.50	3954	0141	0001	0.149	0.128	0.117
	0.30	0.40	2794	1160	0142	0.233	0.201	0.185
	0.20	0.35	0617	2976	0512	0.400	0.347	0.320
Wall	0.40	0.50	3414	0665	0017	0.184	0.104	0.096
	0.30	0.40	1490	1924	0682	0.362	0.209	0.196
	0.20	0.35	0083	2394	1619	0.534	0.319	0.294
Pentagon	0.40	0.50	4008	0088	0000	0.146	0.119	0.105
	0.30	0.40	2866	1142	0088	0.224	0.181	0.179
	0.20	0.35	0564	3084	0448	0.397	0.325	0.324

F-transform compression (i.e. bpp) with edge detection algorithm, intensity based F-transform compression with lossless arithmetic coding (i.e. bpp_A) is also given in the table. Although it is observed in results that the arithmetic coding provides better compression as compared to Huffman coding, but since the presented algorithm supports achieving faster compression at superior quality. Thus, the Huffman code is chosen over the arithmetic code. This result is in line with [29], where it is clearly proved that the Huffman code is having higher performance than arithmetic coding.

Results of the proposed algorithm for achieving compression of images of size 512×512 is shown in Figs. 6 and 7. The corresponding values of these measures (PSNR, RMSE, SAD, SSIM and FSIM) for various images of size 512×512 at different code rates is shown in Table 7. An idea about the time needed during coding

Fig. 6 Compressed images obtained using proposed algorithm (i) row 1 to row3 for block size 16×16 and thresholds (0.4 and 0.5), (0.3 and 0.4) and (0.2 and 0.35) respectively and (ii) row 4 to row 6 for block size 8×8 and thresholds (0.4 and 0.5), (0.3 and 0.4) and (0.2 and 0.35) respectively for the test images of size 512×512

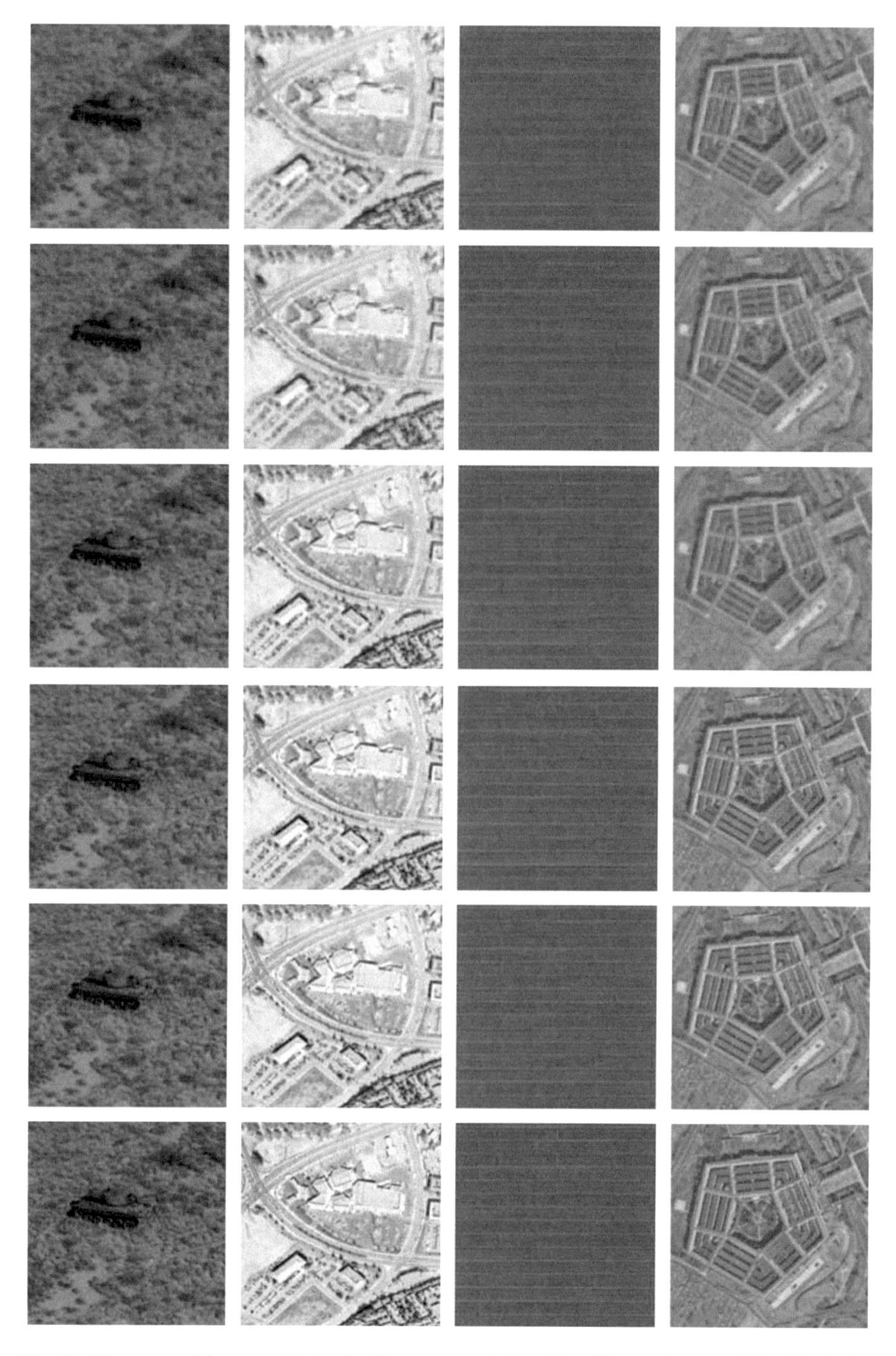

Fig. 7 Compressed images obtained using proposed algorithm (i) row 1 to row3 for block size 16×16 and thresholds (0.4 and 0.5), (0.3 and 0.4) and (0.2 and 0.35) respectively and (ii) row 4 to row 6 for block size 8×8 and thresholds (0.4 and 0.5), (0.3 and 0.4) and (0.2 and 0.35) respectively for the test images of size 512×512

Table 7 Quality parameters obtained from the Proposed algorithm

Images	Code rate	PSNR (dB)	RMSE	SAD	MSSIM	FSIM
Elaine	0.033	26.95	11.45	2010390	0.841	0.886
	0.053	27.10	11.25	1962169	0.860	0.894
	0.092	28.59	9.47	1664300	0.914	0.934
	0.142	30.58	7.53	1395212	0.947	0.965
	0.230	31.03	7.18	1312780	0.957	0.969
	0.378	32.54	6.01	1110832	0.974	0.982
Boat	0.031	23.32	17.39	2852722	0.711	0.806
	0.046	23.53	16.97	2761318	0.743	0.821
	0.085	25.22	13.97	2239399	0.847	0.892
	0.132	26.59	11.93	1909585	0.907	0.937
	0.208	27.51	10.75	1701478	0.931	0.948
	0.357	29.87	8.18	1292754	0.964	0.971
Lake	0.033	22.44	19.23	3152624	0.752	0.818
	0.047	22.59	18.91	3076984	0.773	0.828
	0.091	24.69	14.85	2382606	0.878	0.905
	0.140	26.36	12.26	1979480	0.927	0.945
	0.216	27.02	11.36	1806957	0.942	0.952
	0.366	29.33	8.70	1377865	0.969	0.972
Straw	0.030	18.32	30.94	6520098	0.378	0.656
	0.073	19.51	26.95	5614869	0.613	0.792
	0.104	20.48	24.11	5019019	0.723	0.859
	0.145	21.61	22.31	4609318	0.794	0.899
	0.304	23.28	17.47	3471536	0.896	0.940
	0.458	25.41	13.66	2646641	0.947	0.972
Tank	0.028	26.08	12.64	2528483	0.667	0.799
	0.051	26.76	11.70	2311961	0.749	0.849
	0.081	28.08	10.05	1983966	0.845	0.916
	0.120	28.80	9.25	1821468	0.892	0.944
	0.214	29.94	8.113	1572469	0.926	0.960
	0.359	31.96	6.43	1233007	0.962	0.982
Aerial	0.030	20.44	24.21	4210707	0.588	0.737
	0.042	20.73	23.44	4024164	0.639	0.762
	0.084	22.55	18.93	3166496	0.797	0.870
	0.128	23.73	16.55	2719732	0.870	0.919
	0.201	24.76	14.73	2343067	0.908	0.936
	0.347	27.30	16.99	1684364	0.956	0.968

(continued)

Table 7 (continued)

Images	Code rate	PSNR (dB)	RMSE	SAD	MSSIM	FSIM
Wall	0.019	30.87	7.29	1498038	0.763	0.791
	0.046	31.54	6.75	1385567	0.826	0.846
	0.072	32.31	6.61	1271507	0.881	0.910
	0.104	32.80	5.83	1202858	0.908	0.946
	0.209	33.87	5.16	1037257	0.936	0.956
	0.319	35.47	4.29	856799	0.965	0.982
Pentagon	0.028	24.49	15.19	2830518	0.645	0.779
	0.040	24.77	14.71	2719454	0.690	0.798
	0.077	26.43	12.44	2713948	0.822	0.889
	0.119	27.51	10.73	1891397	0.866	0.936
	0.181	28.38	9.70	1685452	0.914	0.944
	0.325	30.68	7.45	1259609	0.957	0.971

and decoding images of size 512×512 using proposed algorithm, FTR and JPEG for achieving almost similar compression rates for various images is given in Table 8. Comparison of PSNR for different compressed images, achieved using proposed method, RFVQ, FTR, FEQ and JPEG methods of compression with respect to code rate is shown in Fig. 8 for four images of size 256×256 and four images of size 512×512. The increasing curve of the proposed method over other methods shows the superiority of the proposed algorithm. At some bit rate, the RFVQ supersedes proposed edgeFuzzy algorithm but results in higher time complexity because of large number of clusters needed.

In comparison to authors' pairFuzzy [19] algorithm high compression ratio and high PSNR is achieved using proposed algorithm. The use of artifact reduction algorithm reduces the artifacts but at the cost of blurring the compressed image.

6 Conclusion

This chapter presents an edge based image compression algorithm in F-transform domain named edgeFuzzy. Input image blocks are first classified as LI, MI and HI blocks based on the edge image obtained using canny edge detection algorithm. Since LI blocks contain small number of edge pixels and hence less information, is therefore compressed more as compared to MI and HI blocks using F-transform. Huffman encoding is further performed on the compressed image to achieve low bit rate. Both subjective and objective evaluation shows that the proposed algorithm outperforms over other state of art image compression algorithms.

Table 8 Coding and decoding timing for test images (seconds)

Images	Code rate	Proposed coding time	Proposed decoding time	Code rate	FTR coding time	FTR decoding time	Code rate	JPEG coding time	JPEG decoding time
Elaine	0.033	5.45	3.46	0.035	5.71	5.15	0.034	5.63	5.55
	0.053	5.54	2.99	0.062	6.53	5.79	0.061	6.50	6.42
	0.092	11.04	6.11	0.097	7.59	6.02	0.086	6.70	6.43
	0.142	4.99	2.27	0.140	6.76	5.95	0.140	6.53	6.55
	0.230	5.40	3.55	0.250	6.35	6.20	0.240	6.50	5.95
	0.378	9.81	7.75	0.391	8.13	8.95	0.431	6.25	6.11
Boat	0.031	5.05	3.42	0.035	6.10	5.43	0.036	6.39	6.27
	0.046	5.08	3.21	0.062	6.05	5.25	0.063	5.46	5.48
	0.085	11.48	7.90	0.097	6.27	5.32	0.092	6.41	5.91
	0.132	5.25	3.49	0.140	6.79	6.35	0.140	7.01	6.95
	0.208	5.59	3.55	0.250	7.22	7.15	0.241	6.39	6.35
	0.357	14.03	6.99	0.391	7.29	6.44	0.432	6.43	6.25
Lake	0.033	5.84	3.46	0.035	7.53	6.59	0.036	6.47	5.48
	0.047	6.07	3.59	0.062	7.47	6.75	0.062	5.86	5.55
	0.091	8.50	5.13	0.097	6.25	5.49	0.089	5.79	5.65
	0.140	2.95	1.58	0.140	6.27	5.52	0.142	5.89	6.21
	0.216	4.24	1.59	0.250	6.99	6.58	0.239	6.01	6.00
	0.366	6.30	3.43	0.391	7.16	6.53	0.429	6.12	6.31
Straw	0.030	5.73	3.50	0.035	6.84	6.65	0.035	6.61	4.55
	0.073	5.68	3.26	0.062	7.31	6.60	0.062	6.22	6.21
	0.104	10.67	6.14	0.097	6.95	6.95	0.089	6.63	6.13
	0.145	3.23	1.48	0.140	7.62	7.11	0.139	5.89	5.95
	0.304	3.67	1.62	0.250	6.50	6.48	0.249	5.92	5.45
	0.458	11.78	6.38	0391	7.11	7.02	0.439	6.23	5.58

(continued)

Table 8 (continued)

Images	Code rate	Proposed coding time	Proposed decoding time	Code rate	FTR coding time	FTR decoding time	Code rate	JPEG coding time	JPEG decoding time
Tank	0.028	5.60	3.13	0.035	8.31	7.35	0.034	6.33	6.21
	0.051	3.81	1.88	0.062	8.95	7.39	0.062	6.49	6.25
	0.081	7.56	4.53	0.097	9.04	8.24	0.093	6.59	6.14
	0.120	3.19	1.49	0.140	8.22	7.31	0.136	7.01	6.42
	0.214	3.66	1.54	0.250	7.08	7.00	0.241	7.23	6.59
	0.359	4.79	2.20	0.391	7.25	7.52	0.438	6.95	6.53
Aerial	0.030	2.95	1.43	0.035	4.28	4.52	0.035	6.17	4.28
	0.042	3.08	1.50	0.062	6.94	6.18	0.062	6.12	4.65
	0.084	5.60	2.18	0.097	4.51	4.75	0.094	5.92	5.32
	0.128	3.70	3.02	0.140	6.21	6.06	0.139	6.25	5.26
	0.201	4.14	3.18	0.250	4.95	5.47	0.242	5.93	5.23
	0.347	5.99	2.92	0.391	5.32	5.95	0.434	5.93	5.26
Wall	0.019	3.46	1.84	0.035	5.74	5.26	0.035	5.62	5.56
	0.046	4.77	2.62	0.062	6.06	5.94	0.062	5.96	6.12
	0.072	5.20	1.72	0.097	6.42	6.12	0.090	5.86	6.18
	0.104	5.21	2.77	0.140	5.98	5.69	0.140	6.11	6.54
	0.209	5.51	2.74	0.250	6.43	6.12	0.242	6.24	5.89
	0.319	6.94	2.95	0.391	6.92	5.29	0.426	6.13	5.99
Pentagon	0.028	1.67	0.91	0.035	3.95	1.78	0.044	4.98	5.09
	0.040	2.35	1.21	0.062	4.26	3.12	0.065	5.16	5.26
	0.077	2.29	1.02	0.097	4.78	3.26	0.098	5.23	5.32
	0.119	2.62	1.41	0.140	5.56	3.56	0.152	5.26	5.45
	0.181	4.87	2.38	0.250	5.53	3.93	0.301	5.84	5.82
	0.325	5.96	2.97	0.391	4.08	2.96	0.394	5.62	5.71

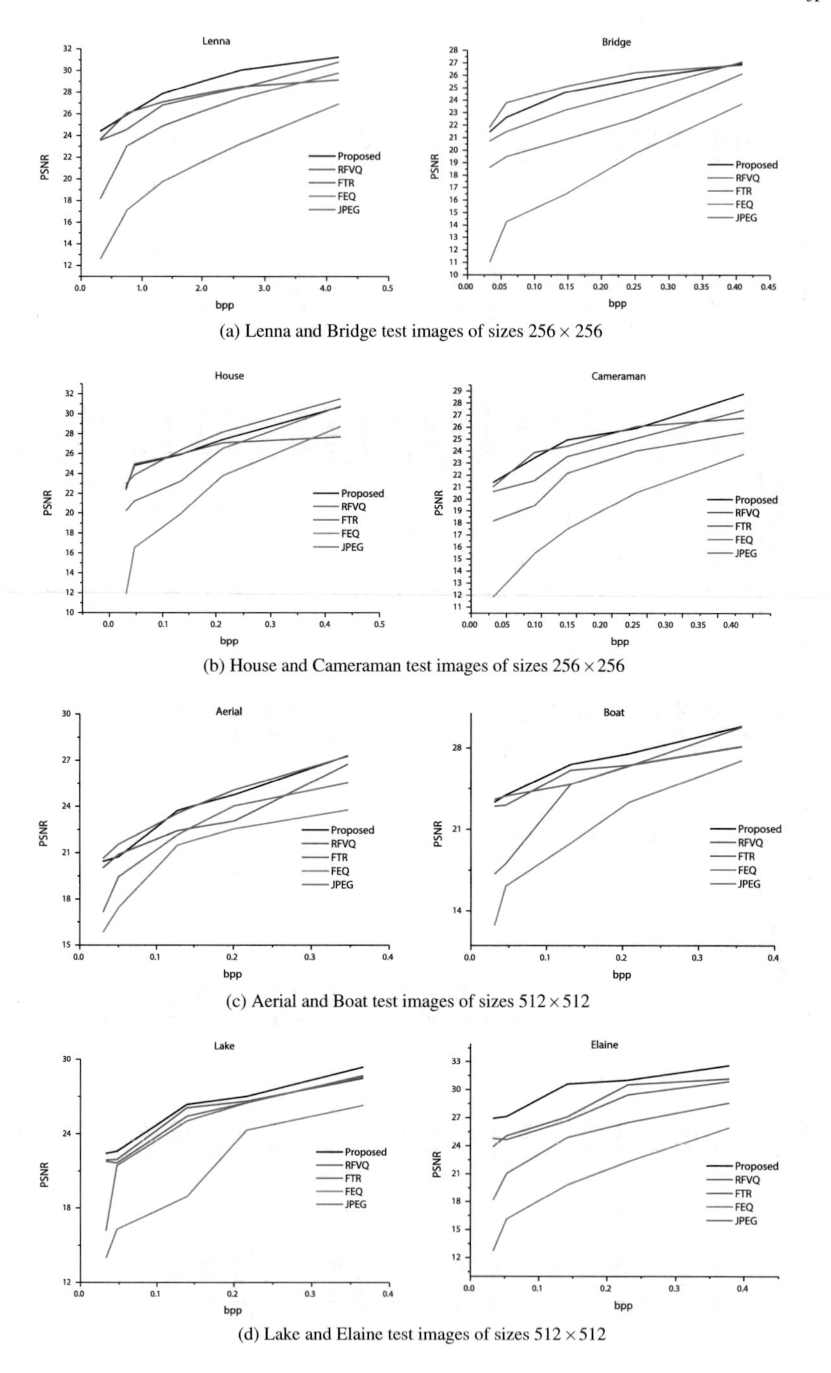

(a) Lenna and Bridge test images of sizes 256 × 256

(b) House and Cameraman test images of sizes 256 × 256

(c) Aerial and Boat test images of sizes 512 × 512

(d) Lake and Elaine test images of sizes 512 × 512

Fig. 8 PSNR comparison of Proposed, RFVQ, FTR, FEQ and JPEG methods

References

1. Lossy image compression, in *Encyclopedia of GIS* (Springer, US, 2008)
2. M. Biswas, S. Kumar, T.Q. Nguyen, N. Balram, Support vector machine (SVM) based compression artifact-reduction technique. J. Soc. Inf. Display **15**(8), 625–634 (2007)
3. S. Saha, Image compression: from DCT to wavelets: a review. ACM Crossroads **6**(3), 12–21 (2000)
4. S. Wang, T. Lin, United coding method for compound image compression. Multimedia Tools Appl. **71**(3), 1263–1282 (2014)
5. W. Xiaolin, X. Zhang, X. Wang, Low bit-rate image compression via adaptive down-sampling and constrained least squares upconversion. IEEE Trans. Image Process. **18**(3), 552–561 (2009)
6. L. Xi, L. Zhang, A study of fractal image compression based on an improved genetic algorithm. Int. J. Nonlinear Sci. **3**(2), 116–124 (2007)
7. Q. Xia, X. Li, L. Zhuo, K.M. Lam, A novel low-bit-rate image compression algorithm, in *Advances in Multimedia Information Processing-PCM 2010* (Springer, 2011), pp. 100–110
8. I. Perfilieva, B. De Baets, Fuzzy transforms of monotone functions with application to image compression. Inf. Sci. **180**(17), 3304–3315 (2010)
9. F. Di Martino, V. Loia, I. Perfilieva, S. Sessa, An image coding/decoding method based on direct and inverse fuzzy transforms. Int. J. Approx. Reason. **48**(1), 110–131 (2008)
10. A. Petrosino, A. Ferone, Rough fuzzy set-based image compression. Fuzzy Sets Syst. **160**(10), 1485–1506 (2009)
11. L. Wang, L. Jiao, W. Jiaji, G. Shi, Y. Gong, Lossy-to-lossless image compression based on multiplier-less reversible integer time domain lapped transform. Signal Process.: Image Commun. **25**(8), 622–632 (2010)
12. D. Ke, L. Peng, New algorithms for preserving edges in low-bit-rate wavelet-based image compression. IEEJ Trans. Electr. Electron. Eng. **7**(6), 539–545 (2012)
13. U. Desai, I. Masaki, A. Chandrakasan, B.K.P. Horn, Edge and mean based image compression, in *Proceedings of IEEE Acoustics, Speech, and Signal Processing, ICASP 1996*, vol. 49 (1996)
14. A. Mertins, Image compression via edge-based wavelet transform. Opt. Eng. **38**(6), 991–1000 (1999)
15. A. Avramovic, Lossless compression of medical images based on gradient edge detection, in *Proceedings of the 19th Telecommunications Forum (TELFOR), Belgrade* (2011), pp. 1199–1202
16. D. Gambhir, N. Rajpal, Fuzzy edge detector based adaptive quantization image coding: FuzzAQC, in *Proceedings of the Recent Advances in Information Technology (RAIT)* (2012), pp. 101–106
17. T.M. Amarunnishad, V.K. Govindan, A.T. Mathew, Improving BTC image compression using a fuzzy complement edge operator. Signal Process. **88**(12), 2989–2997 (2008)
18. D. Gambhir, N. Rajpal, Image coding using fuzzy edge classifier and fuzzy f-transform: dualFuzzy. Int. J. Fuzzy Comput. Modell. **1**(3), 235–251 (2015)
19. D. Gambhir, N. Rajpal, Improved fuzzy transform based image compression and fuzzy median filter based its artifact reduction: pairfuzzy. Int. J. Mach. Learn. Cybern. **6**(6), 935–952 (2015)
20. I. Perfilieva, Fuzzy transforms. Fuzzy Sets Syst. **15**(8), 993–1023 (2006)
21. I. Perfilieva, R. Valásek, Data compression on the basis of fuzzy transforms, in *EUSFLAT Conference* (Citeseer, 2005), pp. 663–668
22. J. Canny, A computational approach to edge detection. IEEE Trans. Pattern Anal. Mach. Intell. **6**, 679–698 (1986)
23. D.A Huffman et al., A method for the construction of minimum redundancy codes, in *Proceedings of I.R.E* (1952), pp. 1099–1101
24. USC-SIPI image database. http://sipi.usc.edu/database/
25. Z. Wang, A.C. Bovik, Mean squared error: love it or leave it? a new look at signal fidelity measures. IEEE Signal Process. Mag. **26**(1), 98–117 (2009)
26. Z. Wang, A.C. Bovik, H.R. Sheikh, E.P. Simoncelli, Image quality assessment: from error visibility to structural similarity. IEEE Trans. Image Process. **13**(4), 600–612 (2004)

27. L. Zhang, D. Zhang, X. Mou, FSIM: a feature similarity index for image quality assessment. IEEE Trans. Image Process. **20**(8), 2378–2386 (2011)
28. S. Serikawa, H. Lu, L. Zhang, Maximum local energy: an effective approach for image fusion in beyond wavelet transform domain. Comput. Math. Appl. **64**(5), 996–1003 (2012)
29. A. Shahbahrami, R. Bahrampour, M. Sabbaghi Rostami, M.A. Mobarhan, Evaluation of huffman and arithmetic algorithms for multimedia compression standards. arXiv:1109.0216 (2011)

Real-Time Implementation of Human Action Recognition System Based on Motion Analysis

Kamal Sehairi, Cherrad Benbouchama, El Houari Kobzili and Fatima Chouireb

Abstract This paper proposes a Pixel Streams-based FPGA implementation of a real-time system that can detect and recognize human activity using Handel-C. The first part of this work details the hardware implementation of a real-time video surveillance system on FPGA, including all the stages, i.e., capture, processing, and display, using DK IDE. The targeted circuit is an XC2V1000 FPGA embedded on Agility's RC200E board. In the second part of this work, we propose a GUI programmed using Visual C++ to facilitate the implementation for novice users. Using this GUI, the user can program/erase the FPGA or change the parameters of different algorithms and filters. The PixelStreams-based implementation was successfully realized and validated for real-time motion detection and recognition.

Keywords Detection · Recognition · FPGA · Real-time implementation · Video surveillance

1 Introduction

In modern society, there is a growing need for technologies such as video surveillance and access control to detect and identify human and vehicle motion in various situations. Intelligent video surveillance attempts to assist human operators when the number of cameras exceeds the operators' capability to monitor them and alerts the operators when abnormal activity is detected. Most intelligent video surveillance systems are designed to detect and recognize human activity. It is difficult to define abnormal activity because there are many behaviors that can

K. Sehairi (✉) · F. Chouireb
Laboratoire LTSS, Université Amar Telidji Laghouat, route de Ghardaia, 03000 Laghouat, Algeria
e-mail: sehairikamel@yahoo.fr; k.sehairi@lagh-univ.dz

C. Benbouchama · E.H. Kobzili
Laboratoire LMR, École Militaire Polytechnique, Algiers, Algeria

H. Lu and Y. Li (eds.), *Artificial Intelligence and Computer Vision*,
Studies in Computational Intelligence 672, DOI 10.1007/978-3-319-46245-5_9

represent such activity. Examples include a person entering a subway channel, abandonment of a package, a car running in the opposite direction, and people fighting or rioting. However, it is possible not only to set criteria to detect abnormal activity but also to zoom in on the relevant area to facilitate the work of the operator.

In general, an intelligent video surveillance system has three major stages: detection, classification, and activity recognition [1]. Over the years, various methods have been developed to deal with issues in each stage.

2 Related Work

Many methods for motion detection have already been proposed. They have been classified [1–3] into three major categories: background subtraction [4, 5], temporal differencing [6, 7], and optical flow [8, 9]. Further, motion detection methods have been recently classified into matching methods, energy-based methods, and gradient methods. The aim of the motion detection stage is to detect regions corresponding to moving objects such as vehicles and human beings. It is usually linked to the classification stage in order to identify moving objects. There are two main types of approaches for moving object classification: [1, 2, 10] shape-based identification and motion-based classification. Different descriptions of shape information of motion regions such as representations of points, boxes, silhouettes, and blobs are available for classifying moving objects. For example, Lipton et al. [11]. used the dispersedness and area of image blobs as classification metrics to classify all moving object blobs into human beings, vehicles, and clutter. Further, Ekinci et al. [12]. used silhouette-based shape representation to distinguish humans from other moving objects, and the skeletonization method to recognize actions. In motion-based identification, we are more interested in detecting periodic, non-rigid, articulated human motion. For example, Ran et al. [13]. examined the periodic gait of pedestrians in order to track and classify it. The final stage of surveillance involves behavior understanding and activity recognition. Various techniques for this purpose have been categorized into seven types: dynamic time warping algorithms, finite state machines, hidden Markov models, time-delay neural networks, syntactic techniques, non-deterministic finite automata, and self-organizing neural networks. Such a wide variety of techniques is attributable to the complexity of the problems and the extensive research conducted in this field. The computational complexity of these methods and the massive amount of information obtained from video streams makes it difficult to achieve real-time performance on a general-purpose CPU or DSP. There are four main architectural approaches for overcoming this challenge: application-specific integrated circuits (ASICs) and field-programmable gate arrays (FPGAs), parallel computing, GPUs, and multiprocessor architectures. Evolving high-density FPGA architectures, such as those with embedded multipliers, memory blocks, and high I/O (input/output) pin counts, are ideal solutions for video processing applications [14]. In the field of image and video processing, there are many FPGA implementations for motion segmentation and tracking. For example, Menezes et al. [5] used background subtraction to detect vehicles in motion, targeting Altera's Cyclone II FPGA with Quartus II software. Another similar study on road traffic detection [15] adopted the sum of absolute differences (SAD) algorithm, implemented on Agility's RC300E board using an XC2V6000 FPGA with Handel-C and the PixelStreams library of Agility's DK Design Suite. Other methods for motion detection such as optical flow have been successfully implemented [8, 9] on an FPGA. For example, Ishii et al. [8] optimized an optical flow algorithm to process 1000 frames per second. The algorithm was implemented on a Virtex-II Pro FPGA.

Many video surveillance systems have been developed for behavior change detection. For example, in the framework of ADVISOR, a video surveillance system for metro stations, a finite state machine (with scenarios) [16] is used to define suspicious behavior (jumping over a barrier, overcrowding, fighting, etc.). The W4 system [17] is a system for human activity recognition that has been implemented on parallel processors with a resolution of 320 $\times$ 240. This system can detect objects carried by people and track body parts using background detection and silhouettes. Bremond and Morioni [18] extracted the features of moving vehicles to detect their behaviors by setting various scenario states (toward an endpoint, stop point, change in direction, etc.); the application employs aerial grayscale images.

The objective of this study is to implement different applications of behavior change detection and moving object recognition based on motion analysis and the parameters of moving objects. Such applications include velocity change detection, direction change detection, and posture change detection. The results can be displayed in the RGB format using chains of parallelized sub-blocks. We used Handel-C and the PixelStreams library of Agility's DK Design Suite to simplify the acquisition and display stages. An RC200E board with an embedded Virtex-II XC2V1000 FPGA was employed for the implementation.

3 Mixed Software-Hardware Design

To make our implementation more flexible, we use the software-hardware platform approach. This approach simplifies not only the use of the hardware but also the change between soft data and hard data, especially for image processing applications that need many parameters to be changed, for example, the parameters of convolution filters and threshold levels. In our implementation, we use Handel-C for the hardware part. Handel-C is a behavior-oriented programming language for FPGA hardware synthesis, and it is adapted to the co-design concept [19].

The software part is developed using Visual C++. After generating the bit file using Agility's DK Design Suite [20], we use our software interface to load this bit file via the parallel port (with a frequency of 50 MHz) on the RC200E board in order to configure the FPGA. The algorithm parameters are transferred through this port as 8-bit data at the same frequency. For the user, these operations are hidden. The graphical user interface allows the user to configure/erase the FPGA and change the algorithm parameters. For example, in our case, we can change the threshold level according to the brightness of the scene or the velocity level according to the object in motion (human, vehicle).

4 Outline of the Algorithm

4.1 *PixelStreams Library*

Before we detail and explain our algorithm and the method used to achieve our goals, we should discuss the tools used for our implementation. We used an RC200E board with an embedded XC2V1000 FPGA [21]. This board has multiple video inputs (S-video, camera video, and composite video), multiple video outputs (VGA, S-video, and composite video), and two ZBT SRAMs, each with a capacity of 2 MB. The language used is Handel-C [22] and the integrated development environment (IDE) is Agility's DK5. This environment is equipped with different platform development kits (PDKs) that include the PixelStreams library [23].

The PixelStreams library is used to develop systems for image and video processing. It includes many blocks (referred to as filters) that perform primary video processing tasks such as acquisition, stream conversion, and filtering. The user has to associate these blocks carefully by indicating the type of the stream (pixel type, coordinate type, and synchronization type). Then, the user can generate the algorithm in Handel-C. Thereafter, the user has to add or modify blocks to program his/her method, and finally, he/she must merge the results. It is worth mentioning that these blocks are parameterizable, i.e., we can modify the image processing parameters, such as the size of the acquired image or the threshold. These blocks are fully optimized and parallelized. Figure 1 shows the GUI of PixelStreams.

4.2 *Detection Algorithm*

We choose to implement the delta frame method for three reasons: its adaptability to changes in luminance, its simplicity, and its low consumption of hardware resources. This method determines the absolute difference between two successive images, and it is executed in two stages: temporal difference and segmentation.

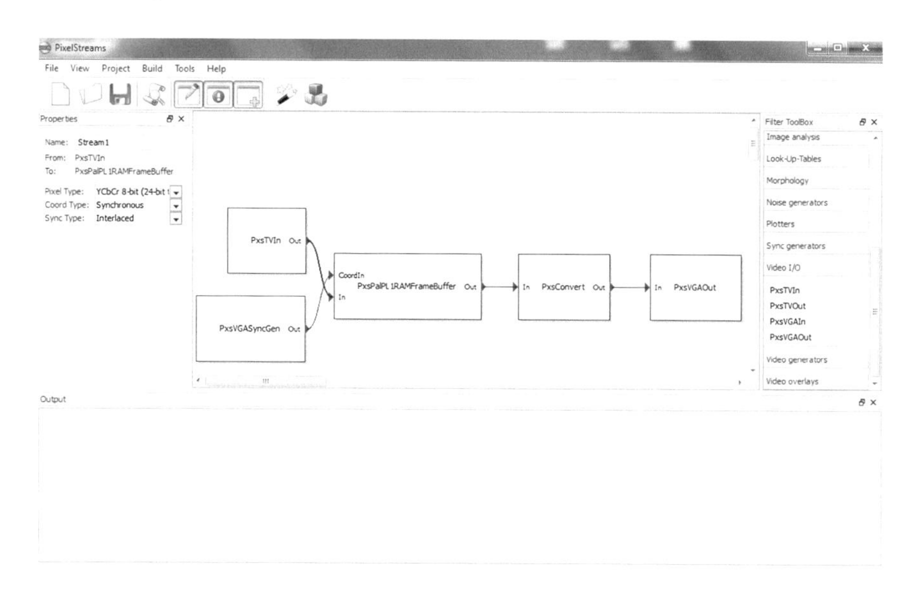

Fig. 1 PixelStreams GUI

4.2.1 Temporal Difference

In this stage, we determine the absolute difference between the previous frame and the current frame as follows.

$$\zeta(x, y) = \left|\frac{dI(x, y)}{dt}\right| = |\Delta_{t,t-1}(x, y)| = |I_t(x, y) - I_{t-1}(x, y)| \tag{1}$$

where $\zeta(x, y)$ is the difference between $I_t(x, y)$ (i.e., the intensity of pixel (x, y) at moment t) and $I_{t-1}(x, y)$ (i.e., the intensity of pixel (x, y) at moment $t - 1$).

4.2.2 Segmentation

In this stage, significant temporal changes are detected by means of thresholding:

$$\Psi(x, y) = \begin{cases} 0 & \text{if } \zeta(x, y) < Th \\ 1 & \text{otherwise} \end{cases} \tag{2}$$

This operation yields a binary card that indicates zones of significant variations in brightness from one image to the other.

4.3 Feature Extraction and Behavior Change Detection

In this study, simple behavior change detection refers to motion that can be caused by abrupt movements that might represent suspicious actions. To define these actions, we use the parameters of the objects in motion, such as the center of gravity, width, and length. In general, the actions detected by this method are simple yet useful in video surveillance. For example, velocity change detection is useful for detecting a criminal who is being chased by the police or a car that exceeds the speed limit; direction change detection is useful for detecting a car that is moving in the wrong direction; and posture change detection is useful for detecting a person who bends to place or pick up an object.

Our implementation involves the following stages: acquisition of the video signal, elimination of noise from the input video signal, detection of moving regions, segmentation for separating the moving objects, extraction of the object parameters, classification of the moving objects, and determining whether movements are suspicious.

4.3.1 Velocity Change Detection

We can detect suspicious behavior of a person from his/her gait as well as his/her change in velocity near sensitive locations such as banks, airports, and shopping centers. In such cases, we can calculate the speed (in pixels/s) or acceleration (in pixels/s^2) of the suspect in the image space in real time. There are several ways of representing this anomaly: the most widely adopted method in the literature is the use of a bounding box (a rectangle around the suspect).

It is easy to calculate the speed of a moving object. As soon as the speed or acceleration of the object exceeds a certain threshold of normality (predetermined experimentally or on the basis of statistical studies), a bounding box appears around the suspect. However, the issue that needs to be addressed is the calculation of the speed in real-time circuits owing to the absence of mathematical functions (such as square root), types of data (integer or real values), and the object parameters on which we base our calculation.

In general, the speed and acceleration are calculated as follows:

$$velocity\,(t) = \left(\sqrt{(x_g(t) - x_g(t-dt))^2 + (y_g(t) - y_g(t-dt))^2}\right)/dt \tag{3}$$

$$acceleration\,(t) = (velocity\,(t) - velocity(t-dt))/dt \tag{4}$$

where $x_g(t)$, $y_g(t)$ and $x_g(t - dt)$, $y_g(t - dt)$ are the co-ordinates of the center of gravity of the object at moments t and $t - dt$, respectively, $dt = 40$ ms in our case, and $velocity\,(t)$ and $velocity\,(t - dt)$ are the velocities of the object at moments t and $t - dt$, respectively.

4.3.2 Direction Change Detection

Changes in direction or motion in the wrong direction can represent abnormal behavior depending on the situation. For example, roaming around a building or car can be considered as an abrupt cyclic change in direction, possibly indicating the intention of burglary or car theft. Other examples include detection of a car that is moving in the wrong direction or a person who is moving in the opposite direction of a queue at an exit gate or exit corridor in an airport.

To determine the direction, we select parameters that distinguish the object of interest, such as its center of gravity, width, and length. In general, the co-ordinates of the center of gravity can be used to determine whether the object has changed its direction, i.e., whether it has moved rightward or leftward depending on the position of the camera.

The change in direction along the x-axis is given by

$$x_g(t) - x_g(t - dt)\begin{cases} > 0 \text{ The object did not change direction} \\ < 0 \text{ The object changed direction} \end{cases} \quad (5)$$

The change in direction along the y-axis is given by

$$y_g(t) - y_g(t - dt)\begin{cases} > 0 \text{ The object did not change direction} \\ < 0 \text{ The object changed direction} \end{cases} \quad (6)$$

These techniques, which are based on the object parameters, can be improved by integrating them with advanced models such as finite state machines (FSMs).

5 Hardware Implementation

Figure 2 shows the general outline of our FPGA implementation.

This general outline consists of four blocks: an acquisition block, an analysis block, a display block, and an intermediate block between the display block and the analysis block.

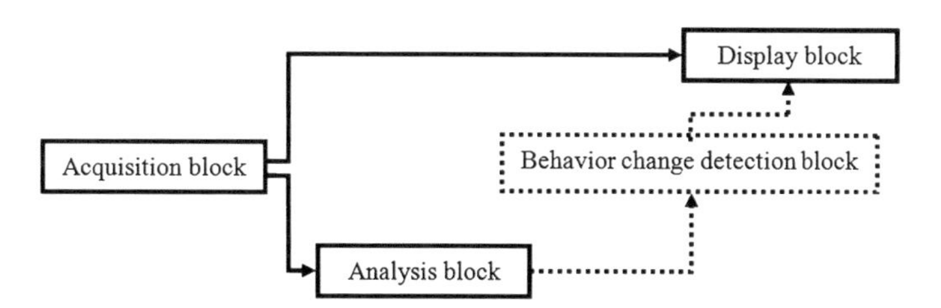

Fig. 2 General outline of behavior change detection

5.1 Acquisition Block

Acquisition is achieved using the standard camera associated with the RC200E board. The video input processor, Philips SAA7113H, acquires the frames in the PAL format at a rate of 25 fps. The pixels are in the YCbCr format. Using the PixelStreams library of Agility's DK Design Suite, we split the input video signal into two identical streams (see Fig. 3). The first stream is fed to the display block and it is converted into the RGB format to display the results on a VGA display. The second stream is fed to the analysis block and it is converted into the grayscale format to reduce (by one-third) the amount of data to be processed.

We can choose to perform the conversion into the RGB format before splitting the input signal and then convert the second stream for the analysis block into the grayscale format. However, this method is not preferable because the conversion from the YCbCr format to the RGB format is approximate. Moreover, in the conversion from the YCbCr format to the grayscale format, the brightness is simply represented by the Y component of the YCbCr format. Further, it is not preferable to approximate the input stream that is fed to the analysis block.

5.2 Analysis Block

The analysis block consists of several stages. In the first stage, we use inter-image subtraction (delta frames) and apply thresholding to detect moving regions.

To obtain the delta frames, we start by splitting the video signal in three channels (see Fig. 4). The first and second channels are used to save the acquired image, creating a delay cell. The image *I(t − 1)* is recorded in the memory. The third channel is used to acquire the actual frame at moment *t*. Then, the two image

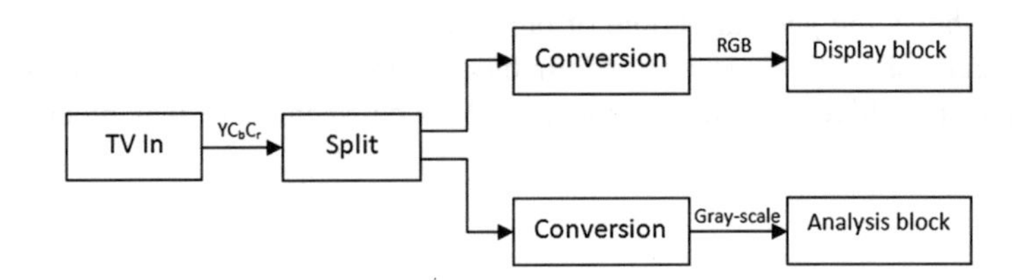

Fig. 3 Acquisition block

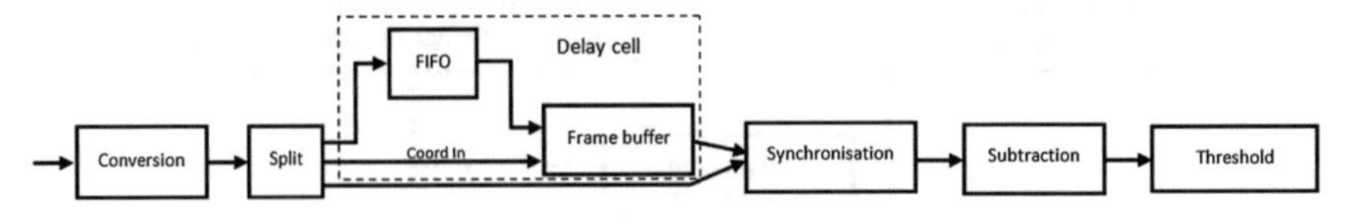

Fig. 4 Motion detection block

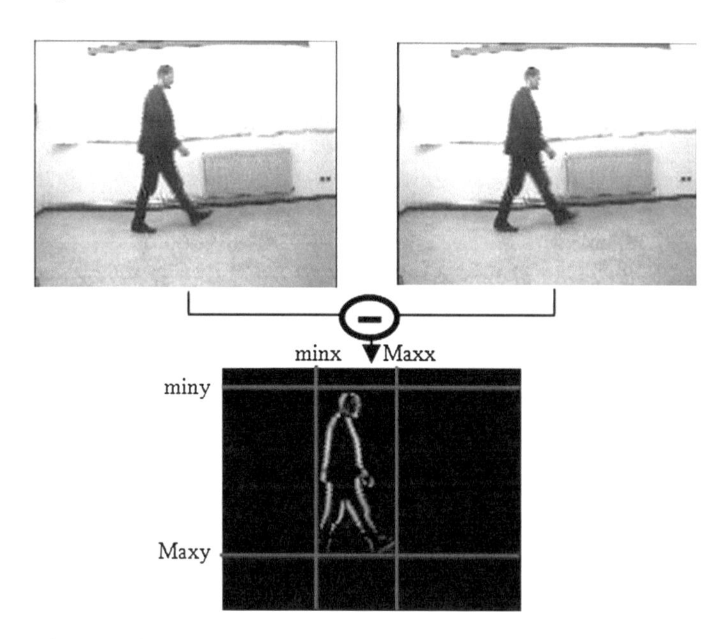

Fig. 5 Inter-image difference and calculation of min and max values

streams are synchronized and fed to the subtraction block. The subtraction block is a modified block that takes the absolute result of subtraction and compares it with a threshold. This function is realized using a macro. The threshold value *Th* is fixed according to the luminosity of the scene.

The second stage of the analysis block involves statistical analysis. In this stage, we search for the min and max values along the x- and y-axes of the mobile regions (Fig. 5). In general, this stage must be preceded by a filter for noise reduction. We employed a morphological filter (e.g., alternating sequential filter, opening/closing filter) using the PixelStreams library.

After calculating the min and max values along the two axes, we determine the center of gravity of the detected object. We calculate the sum of the pixel co-ordinates that have non-zero values along the x- and y-axes, and we divide these coordinate values by their sum. However, for our implementation, it is better to avoid this division. Therefore, we use the direct method. We subtract the max from the min and divide the result by 2. Division by 2 is achieved by a simple bit shift (right shift). Once the values minX, MaxX, minY, MaxY, and X_G, Y_G are obtained, we copy these values into the behavior change detection block. Then, we reset these values to zero.

5.3 Behavior Change Detection Block

As stated in the previous section, the analysis block provides the behavior change detection block with the parameters of the moving objects. In this stage, we save the values extracted from the first delta frame ($x_g(t-1)$, $y_g(t-1)$, $minx(t-1)$,

MaxX(t − 1), miny(t − 1), MaxY(t − 1)), and from the second delta frame, we obtain the current values $x_g(t)$, $y_g(t)$, *minx(t), MaxX(t), miny(t),* and *MaxY(t).* From these latter results, we can calculate the width and length of the moving object to classify the object as human, vehicle, or others, as in our previous work [24]. Using the values extracted in two different instants *(t − 1, t),* we define the changes in behavior.

For velocity change detection, the speed and acceleration are calculated using the two equations presented in Sect. 4.3.1. However, we simplify these equations by calculating the absolute differences between two moments (the previous and current values). If the absolute difference exceeds a certain threshold V_{th}, we assume that the velocity has changed, and we copy the values of the center of gravity in the display block in order to draw a rectangle around the object. Then, the current values are saved as previous values.

Consider a practical problem that involves the values of the center of gravity. In our algorithm, we need to reset all the variables to zero. Consequently, the coordinates of the center of gravity will be zero. If an object enters the scene, the coordinates of the center of gravity change from 0 to X_g, Y_g, and this will cause false detection.

To overcome this problem, we have to ensure that the object has entered the scene entirely. For this purpose, we set a condition on the coordinates of the bounding box for two consecutive instants; if this condition is met (|minXt1 − minXt2 | > S AND |MaxXt1 − MaxXt2 | > S), we can guarantee that the object has entered the scene entirely, either from the right or from the left (Fig. 6 and Table 1).

Figure 7 shows this implementation and represents all the stages realized.

Direction change detection: To implement this application, we follow the same stages as those used in velocity change detection, except that the condition changes. We use the same parameters, *minX, MaxX, minY,* and *MaxY*, in order to avoid the

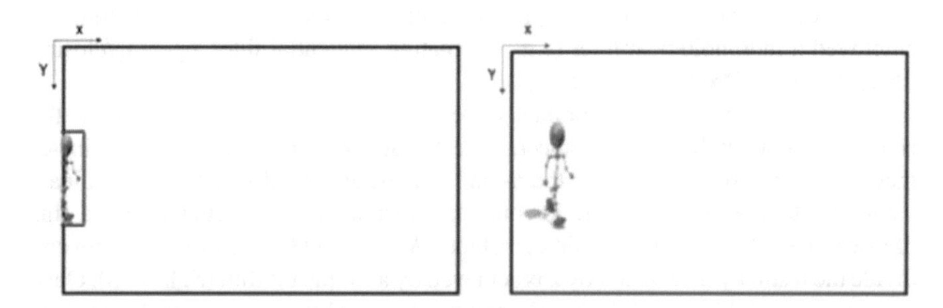

Fig. 6 False velocity change detection (the object enters the scene)

Table 1 Solution proposed for false velocity change detection

$C_1 \Leftrightarrow \lvert \text{min}x(t-1) - \text{min}x(t) \rvert > S$	$C_2 \Leftrightarrow \lvert MaxX(t-1) - MaxX(t) \rvert > S$	$(C_1 \&\& C_2)$
0	0	0
0	1	0
1	0	0
1	1	1

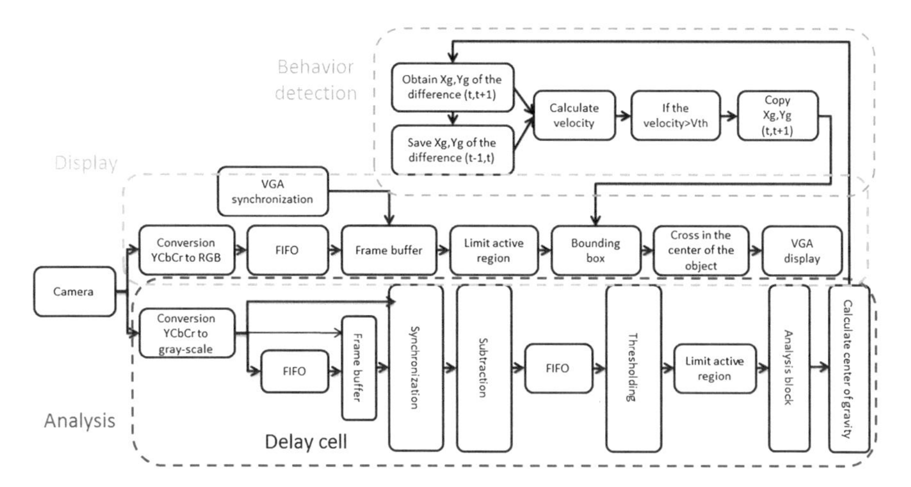

Fig. 7 Hardware architecture for velocity change detection

center of gravity problem. We calculate the difference between minX1 and minX2, and MaxX1 and MaxX2. If there is a change in sign, we assume that the object has changed its direction. Otherwise, we assume that the object has not changed its direction.

We can easily determine the direction of motion of an object by applying the same concept as that described above. However, in this case, it is impractical to compare the differences between the previous values and the current values with zero because the presence of a small or non-significant movement (such as that of the arms) can cause false detection. Therefore, to overcome this problem, we compare the difference with a threshold Th_d, which should not be very large. Then, the values *minX, MaxX, minY,* and *MaxY* are copied to the block that draws the bounding box.

We use two blocks for detection in two directions (a different color for each direction of motion). In order to minimize resource consumption, we used only one block for drawing the bounding box by changing the parameters of entry in our macro. In this macro, we added a parameter that changes the color according to the direction of detected motion (Fig. 8).

Posture change detection: We are interested in such an application to detect a person who leans (bends) to place or pick up something, especially in sensitive locations (e.g., subways). In this case, we are interested in movements along the y-axis of the image (up/down motion), and we use the same architecture as that used in velocity change detection. We calculate the difference between the previous and current values of *miny(t − 1), MaxY(t − 1), miny(t), MaxY(t).*

If the difference between the previous and current values is positive, we assume that the person leans, and we copy the values *minX, MaxX, minY,* and *MaxY* to the block that draws the bounding box and fix the color parameter of the rectangle. We can add a warning message using the PxsConsole filter of PixelStreams. In the

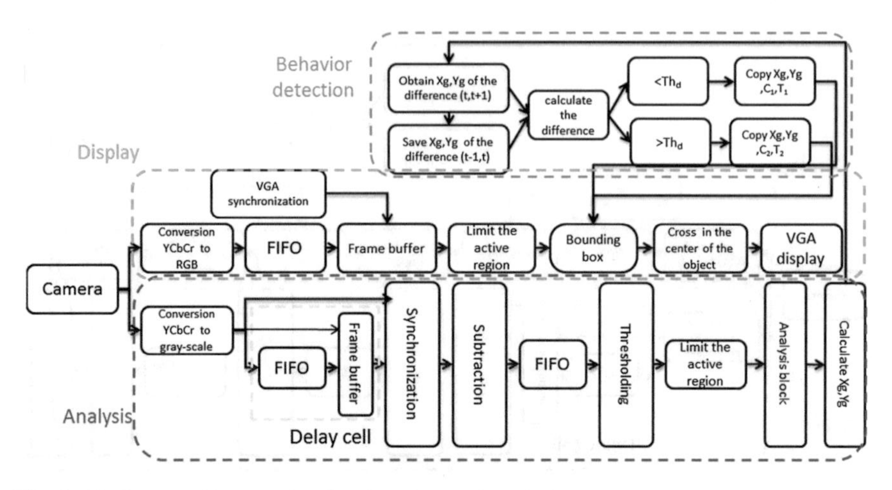

Fig. 8 Hardware architecture for direction change detection

opposite case, we assume that the person rises, and we copy the values to the block that draws the rectangle, which uses a different color in this case. As in the case of direction change detection, it is better to use a threshold Th_p to reduce the occurrence of false detection due to small movements along the y-axis. For such detections, we require a camera whose front sight faces the scene.

Motion analysis: Here, we tried to collect all the above-mentioned behaviors using a single program in order to practically validate the system. To minimize resource consumption, we considered our problem as a finite state machine with several scenarios. The thresholds of detection for each case were used to define and manage these various scenarios. The differences between the values of *minX, MaxX, minY,* and *MaxY* at moments *t* and *t − 1* are denoted by *minx, MaxX, miny,* and *MaxY*, respectively.

In the first state, all the values are initialized (State 0); they represent the initial state of each new inter-image difference. In the second state (State 1), if the ***absolute*** values of *minx and MaxX* are higher than V_{Th}, we assume that the velocity changes and we return to the initial state after copying the values of the block to the bounding box filter. In the opposite case, we go to the third state (State 2) and compare *minx and MaxX* with the threshold Th_d. According to the result of this comparison, we assume that a leftward or rightward movement has occurred. Then, we return to the initial state. Starting from this state, if the moving object accelerates, we return to the second state of velocity change. For posture change detection, the condition is related to the values of *miny* and *MaxY* (State 3). We can detect this behavior from any state (e.g., a person runs and leans to collect something). The following figure summarizes these states and the possible scenarios (Fig. 9).

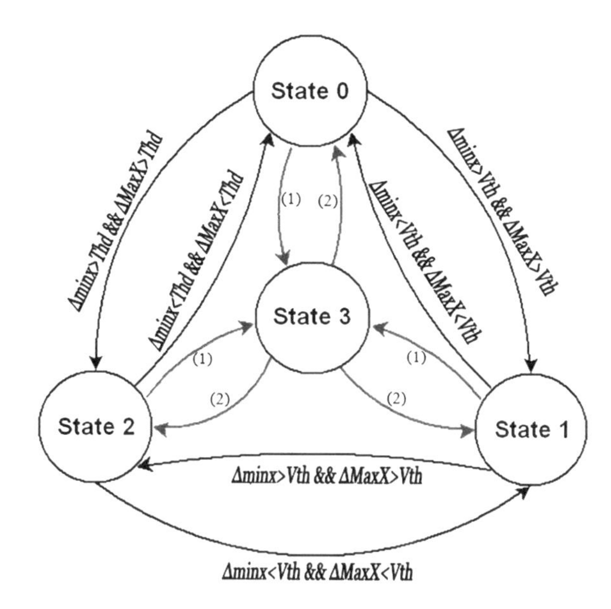

Fig. 9 FSM of motion analysis

5.4 *Display Block*

In this block, we call the macro PxsAnalyseAwaitUpdate, which allows us to pause the display until an update occurs in the analysis block. We obtain the values *minX*, *MaxX*, *minY*, and *MaxY*; if there is a motion, we copy these values to the bounding box filter to draw the rectangle. The values of the center of gravity, X_g, Y_g, are also copied to the PxsCursor filter in order to draw a cross at the center of the moving object. We can add a warning message, e.g., "Warning: velocity change detection", by using the PxsConsole filter of the PixelStreams library. Finally, the results are displayed in the RGB format on a VGA display.

6 Experimental Results

An RC200E board with an embedded Virtex-II XC2V1000 FPGA was used for our implementation. The language used was Handel-C. The results for each behavior are summarized in Tables 2, 3, 4 and 5.

These tables specify the resource consumption and maximal frequency of each implemented detection case for PAL video with a resolution of 720 $\times$ 576.

In all these implementations, the results show that the two main constraints, i.e., the resource limit of our FPGA and the real-time aspect (40 ms/image), are well respected. We note that the consumption of the CLB blocks increases in the case of detection of multiple objects; this is caused by the algorithm used to identify the

Table 2 Resource consumption and maximum frequency of implementation for velocity change detection

Resources	Total	One object	Two objects
I/O	324	179 (55 %)	179 (55 %)
LUTs	10240	2286 (22 %)	3484 (34 %)
Slice Flip/Flops	10240	3046 (29 %)	3738 (36 %)
CLB slices	5120	3092 (60 %)	4040 (78 %)
Block RAM	40	9 (22 %)	9 (22 %)
Frequency	/	67.21 MHz 6.17 ms/image	56.85 MHz 7.29 ms/image

Table 3 Resource consumption and maximum frequency of implementation for direction change detection

Resources	Total	Que object	Two objects
I/O	324	179 (55 %)	179 (55 %)
LUTs	10240	2052 (20 %)	2991 (29 %)
Slice Flip/Flops	10240	2908 (28 %)	3491 (34 %)
CLB slices	5120	2895 (56 %)	3698 (72 %)
Block RAM	40	9 (22 %)	9 (22 %)
Frequency	/	66.69 MHz 6.22 ms/image	55.12 MHz 7.26 ms/image

Table 4 Resource consumption and maximum frequency of implementation for up/down motion detection

Resources	Total	One object	Two objects
I/O	324	179 (55 %)	179 (55 %)
LUTs	10240	2100 (20 %)	3233 (31 %)
Slice Flip/Flops	10240	2927 (28 %)	3595 (35 %)
CLB slices	5120	2961 (57 %)	3904 (76 %)
Block RAM	40	9 (22 %)	9 (22 %)
Frequency	/	67.*14* MHz 6.17 ms/image	58.97 MHz 7.03 ms/image

Table 5 Resource consumption and maximum frequency of implementation for motion analysis

Resources	Total	Two objects
I/O	324	179 (55 %)
LUTs	10240	3640 (35 %)
Slice Flip/Flops	10240	4173 (40 %)
CLB slices	5120	4537 (88 %)
Block RAM	40	9 (22 %)
Frequency	/	53.18 MHz 7.80 ms/image

number of objects in the scene. We also note that the algorithm for motion analysis that collects all the previous behaviors can be implemented on our FPGA in real time, but it consumes nearly all of the CLB resources (88 %).

The following figures show the results of all these implementations. Each behavior is represented by a different color, and a warning message is added below the scenes.

Figure 10 shows the results of velocity change detection in the case of one object. In Fig. 10a, as soon as the object decreases its speed, the rectangle disappears. In Fig. 10b, as soon as the object starts to run, a rectangle appears around it.

Figure 11 shows the results of velocity change detection in the case of two objects. As soon as the objects start running, a rectangle appears. We note that in the case of occlusion, the algorithm considers both objects as a single object. After the objects separate, two rectangles with different colors appear on them.

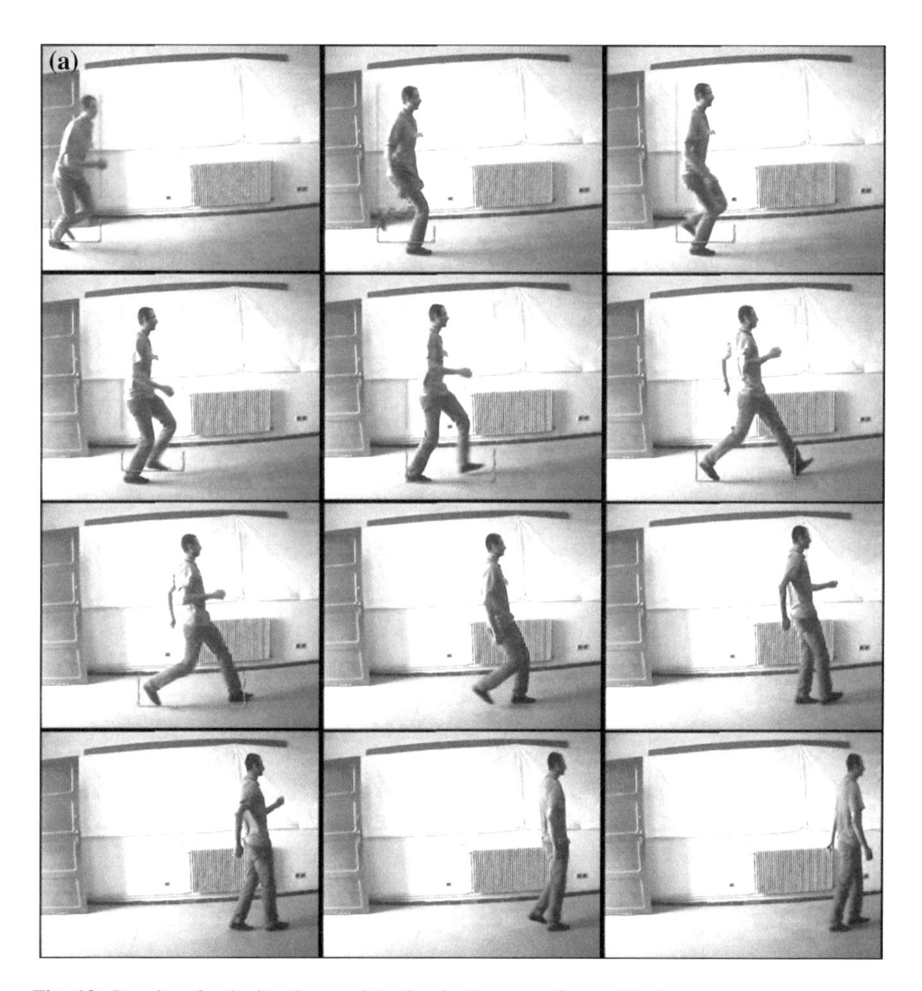

Fig. 10 Results of velocity change detection in the case of one object

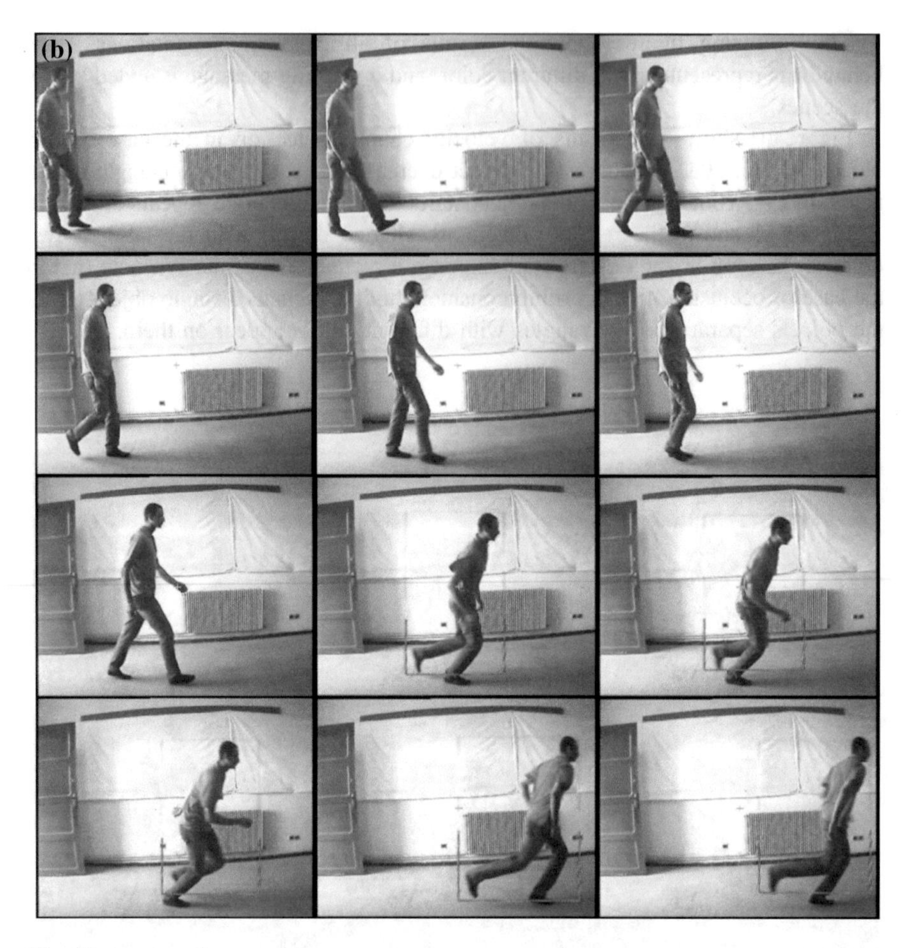

Fig. 10 (continued)

Figure 12 shows the results of direction change detection. Figure 12a shows direction detection for two directions: right to left movement, represented by the blue rectangle, and left to right movement, represented by the red rectangle. The figures also show warning messages below the images. Figure 12b shows the results of direction change detection in one direction for two objects.

Figure 13 shows the results of posture change detection. When the object leans to pick up something, it will be detected. Up/down and down/up motion are represented in different colors. A warning message is added in each case.

Figure 14 shows the results of collecting all the behaviors using a single program. Motion to the right and left are represented by red and blue rectangles,

Fig. 11 Results of velocity change detection in the case of two objects

respectively. Further, up/down and down/up motion are represented by turquoise and yellow rectangles, respectively. Finally, velocity change is represented by a black rectangle. In every case, a warning message is displayed.

Figure 15 shows our graphical user interface (GUI), which is divided into four sections. Three of these sections are used to detect just one simple behavior each, whereas the fourth section detects all the behaviors. Using this GUI, we can send the bit-file for configuring or erasing our FPGA, or directly changing the filter parameters without the need to use the IDE.

Fig. 12 Direction change detection

Fig. 13 Posture change detection, **a** for one object, **b** for two objects

Fig. 14 Motion analysis

7 Conclusion

We presented a mixed software-hardware approach that simplifies the use of the hardware part by enabling us to communicate with it using the graphical interface. In addition, it simplifies the choice of the algorithm to be implemented and modifies the parameters of this algorithm. We adopted the proposed approach for object detection and behavior recognition based on motion analysis and sudden movements. We exploited the hardware part, which offers the possibility of handling large amounts of data and performing calculations for image processing via parallel processing, guaranteed by the use of the PixelStreams library of Agility's DK Design Suite. Further, we tried to improve our architecture by collecting all the different behaviors using a single program. In addition, we added warning messages

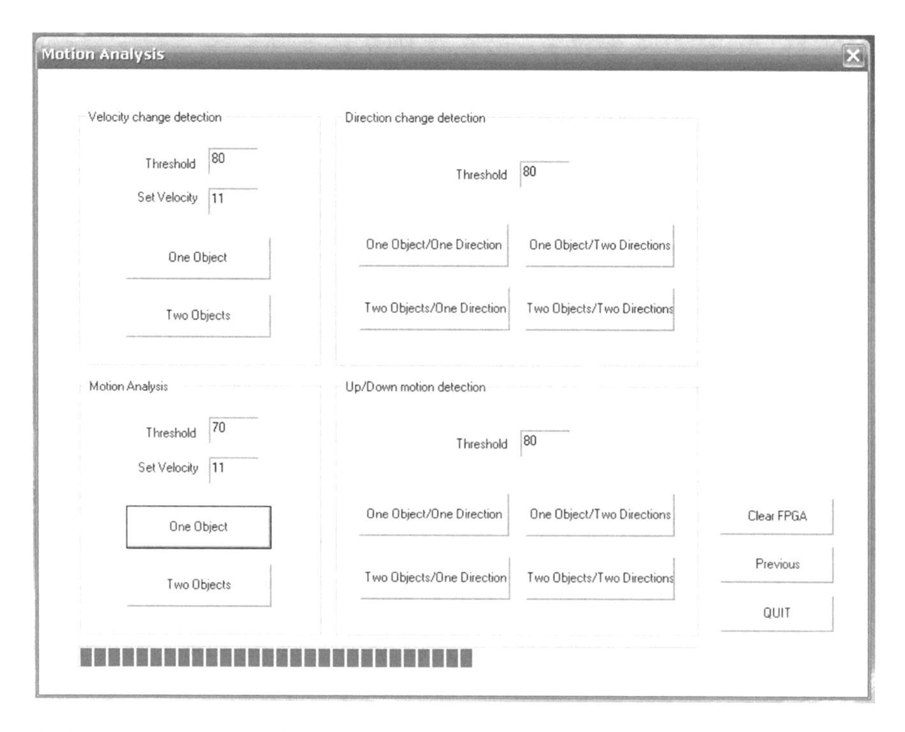

Fig. 15 Graphical user interface

using the PxsConsole filter. Thus, we successfully implemented different algorithms that can recognize objects in motion and detect changes in velocity, direction, and posture in real time. The results showed that our approach achieves good recognition and detection of these behaviors, especially in indoor areas. However, in outdoor areas, the results are less promising owing to the simple motion detection algorithm used; this problem is aggravated by occlusion due to overlapping movements of different persons. Therefore, in the future, we will try to use different and multiple cameras (thermal, infra-red, stereoscopic, Kinect) with improved motion detection and learning methods to detect behavior changes in crowded environments.

Acknowledgments We would like to thank Dr. Benkouider Fatiha for her insightful comments.

References

1. L. Wang, W. Hu, T. Tan, Recent developments in human motion analysis. Pattern Recogn. **36** (3), 585–601 (2003)
2. W. Hu, T. Tan, L. Wang, S. Maybank, A survey on visual surveillance of object motion and behaviors. IEEE T Syst. Man Cybern. **34**(3) (2004)

3. T. Ko, A survey on behavior analysis in video surveillance for homeland security applications, in *AIPR 2008*, Washington, DC, USA (2008)
4. M. Piccardi, Background subtraction techniques: a review. IEEE SMC **4**, 3099–3104 (2004)
5. G.G.S. Menezes, A.G. Silva-Filho, Motion detection of vehicles based on FPGA, in *SPL VI Southern* (2010), pp. 151–154
6. W. Shuigen, C. Zhen, L. Ming, Z. Liang, An improved method of motion detection based on temporal difference. ISA **2009**, 1–4 (2009)
7. M.I.Z. Widyawan, L.E. Nugroho, Adaptive motion detection algorithm using frame differences and dynamic template matching method, in *URAI 2012* (2012), pp. 236–239
8. I. Ishii, T. Taniguchi, K. Yamamoto, T. Takaki, 1000 fps real-time optical flow detection system. Proc. SPIE **7538**, 75380M (2010)
9. J. Diaz, E. Ros, F. Pelayo, E.M. Ortigosa, S. Mota, FPGA-based real-time optical-flow system. IEEE T Circ. Syst. Video **16**(2), 274–279 (2006)
10. M. Paul, S. Haque, S. Chakraborty, *Human Detection in Surveillance Videos and Its Applications—A Review* (Springer, EURASIP JASP, 2013)
11. A.J. Lipton, H. Fujiyoshi, R.S. Patil, Moving target classification and tracking from real-time video, in *WACV 98* (1998), pp. 8–14
12. M. Ekinci, E. Gedikli, Silhouette based human motion detection and analysis for real-time automated video surveillance. Turk. J. Elec. Eng. Comput. Sci. **13**, 199–229 (2005)
13. Y. Ran, I. Weiss, Q. Zheng, L.S. Davis, Pedestrian detection via periodic motion analysis. Int. J. Comput. Vis. **71**(2), 143–160 (2007)
14. K. Ratnayake, A. Amer, An FPGA-based implementation of spatio-temporal object segmentation, in *Proc. ICIP* (2006), pp. 3265–3268
15. M. Gorgon, P. Pawlik, M. Jablonski, J. Przybylo, FPGA-based road traffic videodetector, in *DSD 2007*
16. F. Cupillard, A. Avanzi, F. Bremond, M. Thonnat, Video understanding for metro surveillance, in *ICNSC 2004*
17. I. Haritaoglu, D. Harwood, L.S. Davis, W4: real-time surveillance of people and their activities. IEEE T Pattern Anal. **22**(8), 809–830 (2000)
18. F. Bremond, G. Medioni, Scenario recognition in airborne video imagery, in *IUW 1998* (1998), pp. 211–216
19. M. Edwards, B. Fozard, Rapid prototyping of mixed hardware and software systems, in *DSD 2002* (2002), pp. 118–125
20. Agility DK user manual. Mentor Graphics Agility (2012). http://www.mentor.com/products/fpga/handel-c/dk-design-suite/
21. Virtex II 1.5v Field-Programmable Gate Arrays. Data sheet, Xilinx Corporation (2001)
22. DK5 Handel-C language reference manual, Agility 2007
23. PixelStreams manual. Mentor Graphics Agility (2012). http://www.mentor.com/products/fpga/handel-c/pixelstreams/
24. K. Sehairi, C. Benbouchama, F. Chouireb, Real time implementation on FPGA of moving objects detection and classification. Int. J. Circ. Syst. Signal Process. **9**, 160–167 (2015)

Cross-Modal Learning with Images, Texts and Their Semantics

Xing Xu

Abstract Nowadays massive amount of images and texts has been emerging on the Internet, arousing the demand of effective cross-modal retrieval. To eliminate the heterogeneity between the modalities of images and texts, the existing subspace learning methods try to learn a common latent subspace under which cross-modal matching can be performed. However, these methods usually require fully paired samples (images with corresponding texts) and also ignore the class label information along with the paired samples. Indeed, the class label information can reduce the semantic gap between different modalities and explicitly guide the subspace learning procedure. In addition, the large quantities of unpaired samples (images or texts) may provide useful side information to enrich the representations from learned subspace. Thus, in this work we propose a novel model for cross-modal retrieval problem. It consists of (1) a semi-supervised coupled dictionary learning step to generate homogeneously sparse representations for different modalities based on both paired and unpaired samples; (2) a coupled feature mapping step to project the sparse representations of different modalities into a common subspace defined by class label information to perform cross-modal matching. We conducted extensive experiments on three benchmark datasets with fully paired setting, and a large-scale real-world web dataset with partially paired setting. The results well demonstrate the effectiveness and reasonableness of the proposed method in performing cross-modal retrieval tasks.

1 Introduction

Over the last decade, the amount of multimedia data on social websites (e.g., Facebook, Twitter, Flickr and Youtube) is growing exponentially. The multimedia data usually come from different channels and consist of multiple modalities, such as texts, audio, images and videos. The explosion and diversity of these data has signif-

X. Xu (✉)
School of Computer Science and Engineering,
University of Electronic Science and Technology of China, Chengdu, China
e-mail: xing.xu@uestc.edu.cn

H. Lu and Y. Li (eds.), *Artificial Intelligence and Computer Vision*,
Studies in Computational Intelligence 672, DOI 10.1007/978-3-319-46245-5_10

icantly increased the demand of more sophisticated content retrieval technologies. However, most prevailing retrieval methods [1–4] can only apply to a unimodal setting, where the query and retrieved items are with the same modality. Nowadays, the *cross-modal retrieval* problem, which intends to search heterogeneous data across different modalities given a query in any media type, has attracted more attentions and has been actively studied [5–8]. Taking multimedia retrieval as an example, when people search information on a specific topic (e.g., "New York Times Square"), they may expect to receive comprehensive result containing different media types (e.g., some text documents are from blogs, some images are from Flickr and some videos are from Youtube). In this work, we focus on the cross-modal retrieval problem with image modality and text modality, and consider two mutual tasks: (1) "*Text2Img*": given a text query (document or several words), finding the most related images [9–11]; (2) "*Img2Text*": given an image query, finding the words [12–14] or phrases that best describe the image [11].

Indeed, there are several challenges existing in the cross-modal retrieval problem: (1) the images and texts have different representations. For example, images are usually represented using real-valued and dense feature descriptors, whereas texts are represented as discrete sparse word count vectors; (2) the different modalities of visual and textual features cannot be matched directly with each other since they have distinct statistical properties; (3) the high level semantic description of text and the extracted low level visual descriptors lead to semantic gap between the modalities of image and text.

To eliminate the diversities between the different modalities, a number of recent approaches focusing on latent subspace learning have been proposed. One popular category of such methods is Canonical Correlation Analysis (CCA) [15] and its extensions [12, 16–18]. These methods are designed to learn a common subspace, in which the correlations in paired samples can also be well preserved and the projected features of different modalities can be measured directly. As an alternative to CCA, Li et al. [19] proposed Cross-modal Factor Analysis (CFA), an extension to Latent Semantic Indexing [20] that distinguishes features from different modalities. CFA favors coupled patterns with high variations while CCA is more sensitive to highly coupled patterns with low variation. Another classical method is Partial Least Squares (PLS) [21]. In [22], Sharma et al.proposed to use PLS for cross-modal face recognition problems [22]. Other methods for cross-modal problems have also been proposed, such as BLM [23]. A common characteristic of these methods is that they need paired training samples of different modalities. They suffer when handling more common data on the web that are unpaired. Figure 1 shows typical examples of paired and unpaired samples on the web. Though missing one modality, the unpaired samples are still helpful to provide useful side information and to enrich the representations of learned subspace. Kang et al. [24] proposed a consistent feature representation learning framework to handle unpaired samples, where basis matrices of different modalities are jointly learned and a local group-based priori is proposed to better utilize block features. This framework can be extended to unpaired samples by allowing corrupted feature matrices.

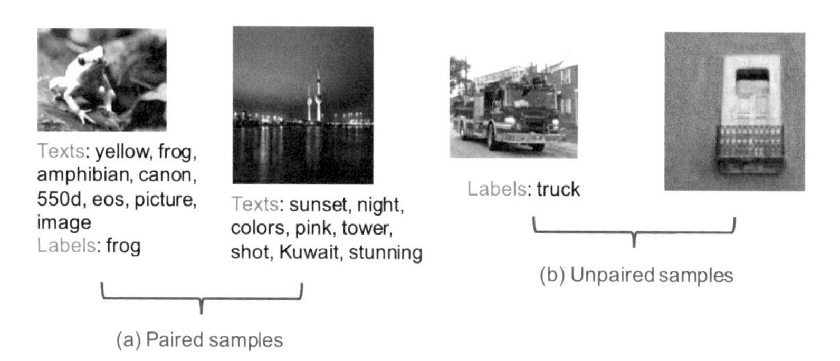

Fig. 1 Examples of paired and unpaired samples. Each paired sample in **a** has image and text with one-to-one correspondences. Each unpaired sample in **b** has image without text. Note that, for both paired and unpaired samples, they can either have ground-truth labels or not

Meanwhile, dictionary learning has become popular for feature representation because of its power in representing heterogeneous features residing in different modalities. Dictionary learning learns a representation space from a set of training samples, where a given signal can be approximately represented as a sparse code. By generating different dictionaries for multi-modal data, dictionary learning can be seamlessly combined with subspace learning and becomes extremely powerful in representing heterogeneous features, this kind of methodology is called coupled dictionary learning (CDL). CDL has recently evolved as a powerful technique and have achieved impressive results in different kinds of tasks, such as face recognition [25], domain adaptation [26] and cross-modal retrieval [27–29]. The main idea of CDL is to learn two dictionaries for the two modalities in a coupled manner such that the sparse coefficients are equal in the original or some transformed space. Like many of the other approaches eg. CCA [15], the standard CDL formulation assumes that the modalities have paired data, i.e. each data point in the first modality is paired with a data point in the second modality. However, it still cannot handle the data coming from the two modalities that are not paired.

Moreover, the aforementioned methods do not make use of the valuable class information to improve model learning while class information can be used to bridge the semantic gap between modalities [30]. As a remedy, Generalized Multiview Analysis (GMA) has been proposed in [31] to exploit for discriminate latent space learning. Kan et al. [32] proposed Multiview Discriminative Analysis (MvDA) for heterogeneous face recognition problems but it increases the dimensions of multi-modal features, making the final matching difficult when the feature dimensions are high.

In conjunction with dictionary learning, Deng et al. [33] propose to learn a discriminative dictionary for each class label and these sub-dictionaries then form the structured discriminative dictionary. The sparse code resulting from this dictionary is then projected into a common label space where cross-modal matching is performed. However, while making use of class label information, these methods once

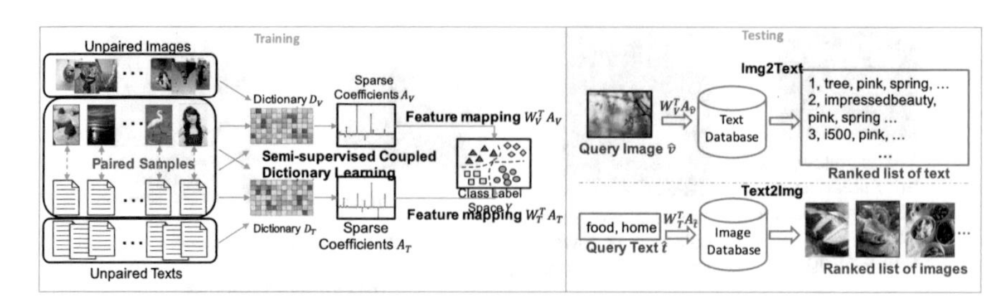

Fig. 2 The overview of our approach. In the training stage, the dictionary for each modality of both paired and unpaired samples are learned in a coupled manner, and the sparse representations of paired samples are simultaneously mapped to the class label space. In the testing stage, the same pipeline is conducted on given query and the cross-modal matching is performed in the class label space

again fail to utilize unpaired data. Kang et al. [24] was able to utilize both unpaired samples and class label information, but their framework was especially optimized for block features, e.g. HOG and GIST, rendering the framework less general.

By now, we can summarize the limitation of these existing methods as following: (1) they only consider the direct correlation between the original representations of images and corresponding text, leading to inefficient latent subspaces for representing data of both modalities; moreover, these methods also have difficulties in discovering the highly non-linear relationship across different modalities; (2) they do not utilize the class label information (e.g., categories or attributes of images) along with the paired samples; (3) they require fully paired samples for training, while ignore the available large amount of unpaired samples. In fact, the class label information would be very helpful to explicitly guide the subspace learning and reduce the semantic gap between the modalities of images and texts. In addition, the unpaired samples can provide useful side information and enrich the representations of learned subspace.

In this work, we focus on (1) learning a more efficient latent subspace from the original representations of different modalities to improve cross-modal retrieval performance and (2) utilizing both paired and unpaired samples simultaneously as well as class information to bridge the semantic gaps. Different from our previous work of [34], here the training stage is a joint framework consisting of two steps: *semi-supervised coupled dictionary learning* and *coupled feature mapping*. We first transform the multi-modal data into sparse representations via coupled dictionary learning, with the guarantee that the generated representations are homogeneous. With the class label information, we derive an efficient feature mapping scheme that projects the learned sparse representations of the paired samples into a discriminant subspace defined by class information. In test stage (i.e. Img2Text or Text2Img), we first generate the sparse representations for the given query from one modality using the learned coupled dictionary of this modality, then map the sparse representations into the common (keyword) subspace using the learned projection matrix. Finally, in the common subspace, we choose the best match from the other modality as the output. Figure 2 visualizes the proposed cross-modal learning framework.

The main contributions of our work can be summarized as follows:

- We propose a novel model that unifies the semi-supervised coupled dictionary learning and coupled feature mapping for the cross-modal retrieval problem. The proposed model can leverage both paired and unpaired samples, as well as the class label information, to improve the efficiency of subspace learning and to boost the retrieval performance.
- We develop an efficient iterative algorithm to solve the complex optimization problem in the proposed model.
- We evaluate the proposed model with both paired and unpaired settings on several challenging datasets and the experimental results show that our model outperforms several relevant state-of-the-art approaches.

The rest of the work is organized as follows. In Sect. 2, we review previous work on multi-modal retrieval. In Sect. 3, we describe the details of the proposed framework and give detailed derivation for the algorithms. Section 4 reports the experimental results on several popular multi-modal datasets and compares the performance with state-of-the-art methods. Finally, we conclude our work with Sect. 5.

2 Related Work

The critical part of multi-modal retrieval is to establish correlation between intrinsically heterogeneous multi-modal representations. The most common method of handling multiple modalities is through learning projections from the feature space of each modality into a common latent space, where features of different modalities become comparable. As a classic example, Canonical Correlation Analysis (CCA) learns a common latent space by maximizing the correlation between the features of two modality [35, 36]. CCA and its extensions have been used in various multi-modal applications. Rasiwasia et al. [17, 37] proposed to use CCA for cross-media retrieval in a two-step framework where CCA is used to learn the maximally correlated subspace. Hwang et al. [38] proposed an unsupervised learning procedure based on Kernel Canonical Correlation Analysis that discovers the relationship between how humans tag images and the relative importance of objects and their layout in the scene. Gong et al. [16] incorporated a third view capturing high-level image semantics, represented either by a single category or multiple non-mutually-exclusive concepts, into CCA for cross-modal retrieval of Internet images and associated text. Zhang et al. [39] proposed a hierarchical subspace learning framework to extract a unique high-level semantic through Isomorphic Relevant Redundant Transformation.

In addition to CCA based approaches, other latent space learning methods for the cross-modal retrieval have been proposed. Chen et al. [40] applied the Partial Least Squares (PLS) to cross-modal document retrieval. They use PLS to transform the visual features into the text space, then learn a semantic space to measure the similarity between two different modalities. Recently, Sharma et al. [31] made a comprehensive analysis for the multi-view learning framework to deal with cross-modal retrieval problem. They extend traditional discriminative methods, i.e., Linear Discriminant Analysis (LDA), Marginal Fisher Analysis (MFA), to the multi-view counterpart: Generalized Multi-view LDA (GMLDA) and Generalized Multi-view MFA (GMMFA). A good feature of this method is that it can incorporate the valuable class label information. Based on the work of Sun et al. [41], Wang et al. [42] proposed a generic framework to jointly perform common subspace learning and coupled feature selection from different modalities of data. They unified coupled linear regression, L_{21} norm and trace norm regularization terms into the generic framework and achieved the state-of-the-art performance for cross-media retrieval tasks. With the development in deep learning, Ngiam et al. [43] proposed a deep Boltzmann machine for cross modality feature learning. Srivastava et al. [5, 44] used deep Boltzmann machine to learn joint space of the image and text for cross-media retrieval. More recently, Peng et al. [45] modeled the cross-modal retrieval problem in terms of graph and proposed a unified graph regularization to optimize the joint feature space; KNN is then used for measuring similarity in the resulting joint feature space.

Combined with subspace learning, dictionary learning is another powerful tool for multi-modal processing. Huang et al. [27] proposed a coupled dictionary learning based model for cross-domain image synthesis and recognition, where a pair of dictionaries for two domains are learned and multi-modal data is then mapped to a common representation space that captures and correlates heterogeneous features. In Shekhar et al. [28] proposed to jointly learn projections in two different domains to construct a discriminative dictionary that can succinctly represent both domains in a projected common low-dimensional representation space.

However, these methods have several drawbacks. First, they only consider the direct correlation between image and text modalities, ignoring the intrinsic diversity of representations and correlation structures in them. This results difficulty in incorporating the highly non-linear relationship between the low-level features across different modalities. Second, most of these methods require fully paired training samples. This renders unpaired or weakly paired data unusable, which largely exist on Internet. Third, most of these methods fails to make good use of the class label information. Class information is very useful in reducing the semantic gaps between different modalities [30].

To overcome these, we develop semi-supervised coupled dictionary learning to handle both paired and unpaired data and integrate class information into the framework to further boost the performance. The sparse representations obtained from dictionary learning are homogeneous for different modalities and can incorporate the relationship across modalities, ensuring learning a more representative latent space. The details of our proposed framework for cross-modal retrieval will be described in the next section.

3 Proposed Framework

In this section, we first show the motivation and problem formulation of the proposed method. We then describe the technical details of the proposed method and explain how to represent and associate cross-modal data by solving semi-supervised coupled dictionary learning and feature mapping in a joint learning framework. Optimization details and complexity study for the training state of the proposed method are subsequently presented. Finally, we present the testing phase of the proposed method for cross-modal retrieval.

3.1 *Problem Formulation*

Let $\boldsymbol{Z} = \{\boldsymbol{Z}^p \cup \boldsymbol{Z}^u\}$ denote a collection of training samples with features from two different modalities. $\boldsymbol{Z}^p = \{(\boldsymbol{v}_i, \boldsymbol{t}_i)\}|_{i=1}^{N^p}$ contains the paired samples, such as the images and their associated texts, where a paired sample $(\boldsymbol{v}_i, \boldsymbol{t}_i)$ consists of d_1 dimensional visual feature $\boldsymbol{v}_i$ and d_2 dimensional text feature $\boldsymbol{t}_i$. $\boldsymbol{V}^p = [\boldsymbol{v}_1, ..., \boldsymbol{v}_{N^p}] \in \mathbb{R}^{d_1 \times N^p}$ and $\boldsymbol{T}^p = [\boldsymbol{t}_1, ..., \boldsymbol{t}_{N^p}] \in \mathbb{R}^{d_2 \times N^p}$ are feature matrices of paired images and texts, respectively. $\boldsymbol{Z}^u = \{(\boldsymbol{v}_j,)|_{j=1}^{N_v^u} \cup (, \boldsymbol{t}_k)|_{k=1}^{N_t^u}\}$ is the subset of N^u unpaired samples including N_v^u images without associated texts and N_t^u texts without corresponding images. Similarly, $\boldsymbol{V}^u \in \mathbb{R}^{d_1 \times N_v^u}$ and $\boldsymbol{T}^u \in \mathbb{R}^{d_2 \times N_t^u}$ are feature matrix of unpaired images and texts, respectively. Specifically, for the top m paired samples in $\boldsymbol{Z}^p$ $(m \leq N^p)$, besides the one-to-one correspondences of visual and text features, they also have class label information $\{\boldsymbol{y}_i\}_{i=1}^{t}$. Here $\boldsymbol{y}_i \in \{0, 1\}^{c \times 1}$ is a c dimensional binary class indicator vector.

For the cross-modal retrieval problems, our primary goal is to learn a representative latent subspace that can get rid of the heterogeneity between different modalities and incorporate the relationships across modalities, by utilizing both paired and unpaired training samples. In addition, an explicit mapping function is also required to ensure the learned space to be discriminant based on the class information of the paired training samples.

3.2 *Semi-supervised Coupled Dictionary Learning*

To handle both paired and unpaired samples of two different modalities, we employ sparse representation from dictionary learning for each modality since it has been shown to be very effective in data representation and reconstruction problems. Specifically, we introduce a semi-supervised coupled dictionary learning method that learns two two dictionaries $\boldsymbol{D}_V \in \mathbb{R}^{d_1 \times k_1}$ and $\boldsymbol{D}_T \in \mathbb{R}^{d_2 \times k_2}$ of two modalities, and the paired samples are used to carry the relationship between different modalities while the unpaired samples are introduced to exploit the marginal distribution for obtaining robust sparse representations. The proposed method can be formulated to minimize the following objective function

$$E_{dl}(\boldsymbol{D}_V, \boldsymbol{D}_T, \boldsymbol{A}_V, \boldsymbol{A}_T) = E^u(\boldsymbol{D}_V, \boldsymbol{A}^u_V) + E^u(\boldsymbol{D}_T, \boldsymbol{A}^u_T) \\ + E^p(\boldsymbol{D}_V, \boldsymbol{D}_T, \boldsymbol{A}^p_V, \boldsymbol{A}^p_T), \quad (1)$$

where $\{\boldsymbol{A}^p_V, \boldsymbol{A}^p_T\}$ and $\{\boldsymbol{A}^u_V, \boldsymbol{A}^u_T\}$ are the sparse coefficients for paired and unpaired samples of two modalities, respectively. Specifically, to guarantee that the sparse representations of the two modalities well reconstruct the unpaired samples, in Eq. 1 we have

$$E^u(\boldsymbol{D}_V, \boldsymbol{A}^u_V) = \|\boldsymbol{V}^u - \boldsymbol{D}_V \boldsymbol{A}^u_V\|_F^2 + \sigma^u_V \|\boldsymbol{A}^u_V\|_1, \\ E^u(\boldsymbol{D}_T, \boldsymbol{A}^u_T) = \|\boldsymbol{T}^u - \boldsymbol{D}_T \boldsymbol{A}^u_T\|_F^2 + \sigma^u_T \|\boldsymbol{A}^u_T\|_1, \quad (2)$$

where $\|\cdot\|_F^2$ is the Frobenius norm for matrices and $\|\cdot\|_1$ is the L_1 norm for constraints of sparsity. Equation 2 is the standard form of sparse coding. And the useful information in unpaired samples can be reflected in the dictionaries of the two modalities. Furthermore, to ensure the dictionaries of two modalities to be coupled, we impose a function $f(\boldsymbol{A}^p_V, \boldsymbol{A}^p_T)$ to relate the sparse representations of two modalities of the paired samples. Then for the paired samples, in Eq. 1, we have

$$E^p(\boldsymbol{D}_V, \boldsymbol{D}_T, \boldsymbol{A}^p_V, \boldsymbol{A}^p_T) = \|\boldsymbol{V}^p - \boldsymbol{D}_V \boldsymbol{A}^p_V\|_F^2 + \|\boldsymbol{T}^p - \boldsymbol{D}_T \boldsymbol{A}^p_T\|_F^2 \\ + \sigma^p(\|\boldsymbol{A}^p_V\|_1 + \|\boldsymbol{A}^p_T\|_1) + f(\boldsymbol{A}^p_V, \boldsymbol{A}^p_T), \quad (3) \\ s.t. \ \|\boldsymbol{d}_{v,i}\|_2 \leq 1, \|\boldsymbol{d}_{t,j}\|_2 \leq 1, \forall i, j.$$

Inspired by [27], we introduce a k_c dimensional common feature space $\mathcal{P}$ for $f(\boldsymbol{A}^p_V, \boldsymbol{A}^p_T)$ so that each pair of samples from $\boldsymbol{A}^p_V$ and $\boldsymbol{A}^p_T$ can be transformed to the points of $\boldsymbol{P}_T$ and $\boldsymbol{P}_V$ in space $\mathcal{P}$ mutually. Here we restrict $k_c = k_1 = k_2$ so that different modalities of data are comparable in $\mathcal{P}$. The formulation of $f(\boldsymbol{A}^p_V, \boldsymbol{A}^p_T)$ is derived as

$$f(\boldsymbol{A}^p_V, \boldsymbol{A}^p_T) = \gamma(\|\boldsymbol{A}^p_V - \boldsymbol{U}_V^{-1} \boldsymbol{P}_T\|_F^2 + \|\boldsymbol{A}^p_T - \boldsymbol{U}_T^{-1} \boldsymbol{P}_V\|_F^2) \\ + \xi(\|\boldsymbol{U}_V^{-1}\|_F^2 + \|\boldsymbol{U}_T^{-1}\|_F^2), \quad (4)$$

where $\boldsymbol{U}_V \in \mathbb{R}^{k_c \times k_1}$ and $\boldsymbol{U}_T \in \mathbb{R}^{k_c \times k_2}$ are the transform matrices for two modalities. The regularized constraints on $\boldsymbol{U}_V$ and $\boldsymbol{U}_T$ ensure numerical stability and avoid overfitting. It can be learned that the constraints in Eq. 4 shows the ability of recovering the sparse representations in one modality using data from the other, hence the relationship across different modalities can be efficiently incorporated in the sparse representations.

3.3 Coupled Feature Mapping

Let $\boldsymbol{Y} = [\boldsymbol{y}_1, \boldsymbol{y}_2, ..., \boldsymbol{y}_n] \in \mathbb{R}^{c\times m}$ be the class label matrix of the paired samples $\{(\boldsymbol{v}_i, \boldsymbol{t}_i)\}|_{i=1}^{m}$. The coupled feature mapping aims to learn two projection matrices $\boldsymbol{W}_V \in \mathbb{R}^{k_1\times c}$ and $\boldsymbol{W}_T \in \mathbb{R}^{k_2\times c}$, which map the sparse representations $\boldsymbol{A}^p_{v_i}|_{i=1}^{m}$ and $\boldsymbol{A}^p_{t_i}|_{i=1}^{m}$ of the two modalities for the paired samples into the common subspace defined by class labels.

In order to minimize the errors of projecting the sparse representations of each modality to the label space, the objective function of the coupled feature mapping procedure can be derived as

$$E_{fm}(\boldsymbol{W}_V, \boldsymbol{W}_T) = \|\boldsymbol{W}_V^{\top}\boldsymbol{A}_V^p - \boldsymbol{Y}\|_F^2 + \|\boldsymbol{W}_T^{\top}\boldsymbol{A}_T^p - \boldsymbol{Y}\|_F^2 + \lambda(\|\boldsymbol{W}_V\|_F^2 + \|\boldsymbol{W}_T\|_F^2). \tag{5}$$

where λ is the regularization parameter. Equation 5 is the standard form of linear classification, which indicates that the sparse representations of the two modalities for the paired samples are mapped into the common label subspace by linear projection.

3.4 Overall Objective Function

Let $\{\boldsymbol{D}_V, \boldsymbol{D}_T, \boldsymbol{A}_V, \boldsymbol{A}_T, \boldsymbol{U}_V, \boldsymbol{U}_T, \boldsymbol{W}_V, \boldsymbol{W}_T\}$ be denoted by $\boldsymbol{\Omega}$, the overall objective function, combining the semi-supervised coupled dictionary step given in Eq. 1 and the coupled feature mapping given in Eq. 5, is formulated as below:

$$\min_{\boldsymbol{\Omega}} E(\boldsymbol{\Omega}) = E_{dl} + \mu E_{fm}, \tag{6}$$
$$s.t. \; \|\boldsymbol{d}_{v,i}\|_2 \le 1, \|\boldsymbol{d}_{t,j}\|_2 \le 1, \forall i, j$$

where $\mu > 0$ controls the combination of the two steps and $\|\cdot\|_2$ is typically applied to avoid trivial solutions.

3.5 Optimization Algorithm

The objective function in Eq. 6 is non-convex with $\boldsymbol{\Omega}$. Fortunately, it is convex with any one in $\boldsymbol{\Omega}$ while fixing the other variables. Therefore, the optimization problem can be solved by an iteratively framework and the variables in $\boldsymbol{\Omega}$ can be updated in an alternating manner until convergency is reached.

3.5.1 Updating D_V and D_T

Learn the dictionaries $\boldsymbol{D}_V$ and $\boldsymbol{D}_T$ by fixing other variables in $\boldsymbol{\Omega}$, the problem in Eq. 1 can be simplified as:

$$\min_{\boldsymbol{D}_V} \|\boldsymbol{V} - \boldsymbol{D}_V\boldsymbol{A}_V\|_F^2, st.\ \|\boldsymbol{d}_{v,i}\|_2 \leq 1, \forall i,$$
$$\min_{\boldsymbol{D}_T} \|\boldsymbol{T} - \boldsymbol{D}_T\boldsymbol{A}_T\|_F^2, st.\ \|\boldsymbol{d}_{t,i}\|_2 \leq 1, \forall i, \tag{7}$$

where $\boldsymbol{V} = [\boldsymbol{V}^u; \boldsymbol{V}^p]$, $\boldsymbol{T} = [\boldsymbol{T}^u; \boldsymbol{T}^p]$, $\boldsymbol{A}_V = [\boldsymbol{A}_V^u; \boldsymbol{A}_V^p]$ and $\boldsymbol{A}_T = [\boldsymbol{A}_T^u; \boldsymbol{A}_T^p]$ include features and sparse representations of both unpaired and paired samples, respectively. Equation 7 is a typical form of quadratically constrained quadratic program (QCQP) with respect to $\boldsymbol{D}_V$ and $\boldsymbol{D}_T$, and it can be efficiently solved using Lagrange dual techniques.

3.5.2 Updating $\{A_V^u, A_T^u\}$ and $\{A_V^p, A_T^p\}$

Similarly, we calculate the solutions of sparse representations $\boldsymbol{A}_V$ and $\boldsymbol{A}_T$ by fixing other variables in $\boldsymbol{\Omega}$.

For $\{\boldsymbol{A}_V^u, \boldsymbol{A}_T^u\}$ of unpaired samples, we have:

$$\min_{\boldsymbol{A}_V^u} \|\boldsymbol{V}^u - \boldsymbol{D}_V\boldsymbol{A}_V^u\|_F^2 + \sigma_V^u\|\boldsymbol{A}_V^u\|_1,$$
$$\min_{\boldsymbol{A}_T^u} \|\boldsymbol{T}^u - \boldsymbol{D}_T\boldsymbol{A}_T^u\|_F^2 + \sigma_T^u\|\boldsymbol{A}_T^u\|_1, \tag{8}$$

which is a form of standard sparse coding form with respect to $\boldsymbol{A}_V^u$ and $\boldsymbol{A}_T^u$.

Similarly, for $\{\boldsymbol{A}_V^p, \boldsymbol{A}_T^p\}$ of paired samples, they can also be formulated as the form of standard sparse coding as:

$$\min_{\boldsymbol{A}_V^p} \|\bar{\boldsymbol{V}} - \bar{\boldsymbol{D}}_V\boldsymbol{A}_V^p\|_F^2 + \sigma^p\|\boldsymbol{A}_V^p\|_1,$$
$$\min_{\boldsymbol{A}_T^p} \|\bar{\boldsymbol{T}} - \bar{\boldsymbol{D}}_T\boldsymbol{A}_T^p\|_F^2 + \sigma^p\|\boldsymbol{A}_T^p\|_1, \tag{9}$$

where $\bar{\boldsymbol{D}}_V = [\boldsymbol{V}^p; \sqrt{\mu}\boldsymbol{W}_V^\top; \sqrt{\gamma}\boldsymbol{U}_V^{-1}\boldsymbol{P}_T]$, $\bar{\boldsymbol{D}}_T = [\boldsymbol{T}^p; \sqrt{\mu}\boldsymbol{W}_T^\top; \sqrt{\gamma}\boldsymbol{U}_T^{-1}\boldsymbol{P}_V]$, $\bar{\boldsymbol{V}} = [\boldsymbol{D}_V; \boldsymbol{Y}; \sqrt{\gamma}\boldsymbol{I}]$ and $\bar{\boldsymbol{T}} = [\boldsymbol{D}_T; \boldsymbol{Y}; \sqrt{\gamma}\boldsymbol{I}]$.

3.5.3 Updating U_V and U_T

Using aforementioned strategy, we can derive the following formulation to update the matrices $\boldsymbol{U}_V$ and $\boldsymbol{U}_T$:

$$\min_{U_V^{-1}} \|A_V^p - U_V^{-1} P_T\|_F^2 + \mu \|U_V^{-1}\|_F^2,$$
$$\min_{U_T^{-1}} \|A_T^p - U_T^{-1} P_V\|_F^2 + \mu \|U_T^{-1}\|_F^2, \tag{10}$$

which are standard ridge regression problems with respect to U_V and U_T. Therefore, we can derive the close-form solutions as:

$$U_V^{-1} = A_V^p P_T^\top [P_T P_T^\top + (\xi/\gamma) I]^{-1},$$
$$U_T^{-1} = A_T^p P_V^\top [P_V P_V^\top + (\xi/\gamma) I]^{-1}, \tag{11}$$

where I is an identity matrix to ensure the issue of full rank during matrix inversion.

3.5.4 Updating W_V and W_T

Similarly as above, regarding W_V and W_T, the problem in Eq. 6 is formulated as:

$$\min_{W_V} \|W_V^\top A_V^p - Y\|_F^2 + \lambda \|W_V\|_F^2,$$
$$\min_{W_T} \|W_T^\top A_T^p - Y\|_F^2 + \lambda \|W_T\|_F^2, \tag{12}$$

which are also standard forms of ridge regression. Therefore, the analytical solutions of W_V and W_T can be derived as:

$$W_V^\top = Y A_V^\top (A_V A_V^\top + \lambda I)^{-1},$$
$$W_T^\top = Y A_T^\top (A_T A_T^\top + \lambda I)^{-1}. \tag{13}$$

The optimization algorithm is summarized in Algorithm 1, and we can iteratively update the variables in $\boldsymbol{\Omega}$ according to the derived solutions above until Eq. 6 is converged.

3.6 *Testing Stage for Cross-Modal Retrieval*

In the testing phase, given a paired sample $(\hat{v}, \hat{t})$, we first generate the sparse representations $\hat{A}_{\hat{v}}$ and $\hat{A}_{\hat{t}}$ based on learned dictionaries D_V and D_T, respectively. Then we can project $\hat{A}_{\hat{v}}$ and $\hat{A}_{\hat{t}}$ into the class label space through the learned projection matrices W_V and W_T, respectively. To perform cross-modal retrieval, we take a sample of one modal data (i.e., an image) as the query to retrieval the other modality of it (i.e., the texts). Even if an unpaired sample with one missing modality is given, we can still reconstruct its missing modality using the similar pipeline.

Algorithm 1 Iterative Algorithm for the proposed method.

Input: Image feature matrices $\{\boldsymbol{V}^p, \boldsymbol{V}^u\}$, text feature matrices $\{\boldsymbol{T}^p, \boldsymbol{T}^u\}$ for paired and unpaired samples, respectively. Class label matrix $\boldsymbol{Y}$ for paired samples. Parameters $\{\sigma^p, \sigma^u_V, \sigma^u_T\}$, γ, μ and λ.

1: Initialize $\{\boldsymbol{D}^0_V, \boldsymbol{D}^0_T\}$ and $\{\boldsymbol{A}^0_V, \boldsymbol{A}^0_T\}$ by [27], and $\{\boldsymbol{U}^0_V, \boldsymbol{U}^0_T\}$ as $\boldsymbol{I}$ for iteration $i = 0$.
2: Let $\boldsymbol{P}^0_V \leftarrow \boldsymbol{U}^0_V \boldsymbol{A}^0_V$ and $\boldsymbol{P}^0_T \leftarrow \boldsymbol{U}^0_T \boldsymbol{A}^0_T$.
3: **repeat**
4: Update $\boldsymbol{D}^{i+1}_V, \boldsymbol{D}^{i+1}_T$ with $\boldsymbol{A}^i_V, \boldsymbol{A}^i_T, \boldsymbol{U}^i_V, \boldsymbol{A}^i_T$.
5: Update $\boldsymbol{A}^{i+1}_V, \boldsymbol{A}^{i+1}_T$ with $\boldsymbol{D}^{i+1}_V, \boldsymbol{D}^{i+1}_T, \boldsymbol{U}^i_V, \boldsymbol{A}^i_T$.
6: Update $\boldsymbol{U}^{i+1}_V, \boldsymbol{U}^{i+1}_T$ with $\boldsymbol{D}^{i+1}_V, \boldsymbol{D}^{i+1}_T, \boldsymbol{A}^{i+1}_V, \boldsymbol{A}^{i+1}_T$.
7: Update $\boldsymbol{P}^{i+1}_V \leftarrow \boldsymbol{U}^{i+1}_V \boldsymbol{A}^{i+1}_V$ and $\boldsymbol{P}^{i+1}_T \leftarrow \boldsymbol{U}^{i+1}_T \boldsymbol{A}^{i+1}_T$.
8: Set $i = i + 1$.
9: **until** Objective function of Eq. 6 converges.

Output: Dictionaries $\{\boldsymbol{D}_V, \boldsymbol{D}_T\}$, sparse representations $\{\boldsymbol{A}_V, \boldsymbol{A}_T\}$ of paired and unpaired samples, mapping matrices $\{\boldsymbol{W}_V, \boldsymbol{W}_T\}$.

3.7 Computational Complexity

The time consuming for the training stage mainly includes sparse coding learning, Lagrange dual learning and ridge regression. Typically, solving Eqs. 9, 7 and 12 requires $\mathcal{O}(N^p d^2)$, $\mathcal{O}(N^u d^2)$ and $\mathcal{O}(d^3)$, respectively. Therefore, the total time complexity of training the proposed method is linear to the number of samples, which is efficient and scalable for large-scale datasets.

4 Experimental Results

In this section, we evaluate the performance of our proposed method for cross-modal retrieval in two settings, *fully paired* setting and *partially paired* setting. Specifically, the *fully paired* setting is the most commonly used setting in cross-modal retrieval problem, in which each image has associated text. The *partially paired* setting, on the other hand, does not guarantee that each image has associated text, i.e. there exist images that are not associated with any text. This is quite common with Internet images, which may have noisy text or do not have any text at all. In the following subsections, we respectively report and discuss results on *fully paired* and *partially paired* settings. For each setting, we first describe the statistics of the datasets, evaluation metrics, and then discuss the results with representative state-of-the-art methods.

Table 1 General statistics of three datasets used for paired setting

Dataset	Images	Labels	Image feature	Text feature
Pascal VOC2007	2808, 2841	20	512-dim Gist	399-dim word frequency
Wiki	1300, 1566	10	128-dim SIFT	10-dim LDA features
MIRFilckr-25K	12500, 12500	38	7500-dim multiple features	457-dim word frequency

4.1 Fully Paired Setting

For *fully paired* setting, we evaluate the proposed method on three publicly available datasets: Pascal VOC2007 [38] and Wiki [17] and MIRFlickr-25K [46], which has been widely used in the cross-modal retrieval field [24, 31, 33]. Each image in the three dataset is associated with corresponding text. Besides, each image is also manually annotated with at least one class label. In particular, each image in Pascal VOC2007 and Wiki datasets is associated with only one label, while in MIRFlickr-25K, each image is assigned with at least one label. Table 1 lists some of the general statistics of the three datasets. We first consider two standard cross-modal retrieval tasks: *Img2Text* and *Text2Img*, for all three datasets. Given an image (or text) query, the goal of each task is to find the nearest neighbors from text (or image) database. In addition, for the multi-label MIRFlickr-25K dataset, we further consider another scenario termed *Img2Label*, which aims to predict proper class labels for an given image query. It is different from the *Img2Text* task and can be treated as the traditional image annotation task. Here we would like to use it to investigate the efficiency of the proposed method on coupled feature mapping into the common label subspace.

The proposed method is compared with several related typical methods, such as PLS [21, 22], CCA [15], GMMFA [31], GMLDA [31], and several recent published state-of-the-art approaches, such as LCFS [42], LGCFL [24] and DLCLA [33]. To evaluate the performance of the *Img2Text* and *Text2Img* tasks, we use the standard measure of mean average precision (MAP) and show the precision scope curve, which are widely adopted in the previous works [24, 31, 33]. To compute MAP, we first evaluate the average precision (AP) of the retrieval result of each query, and then average the AP values over all queries in the query set. To draw the precision-scope curve, we vary the the number (K) of top-ranked samples to a query and compute the AP of all queries correspondingly. For the evaluation of the *Img2Label* task, we adopt the stand measures of image annotation task: average precision per label (P), average recall per label (R). The values of P and R are computed by predicting the top five labels for each test image in MIRFlickr-25K dataset. Note that for all the measures, larger numerical value indicates better performance.

In the training stage, we empirically set the parameters σ, γ and ξ to 0.01, 0.0001 and 0.001 for the coupled dictionary learning procedure; the parameters λ, μ to 0.01 and 1 for the feature mapping procedure, respectively. In testing phase, the cosine distance is adopt to measure the similarity of features and select the matches.

4.1.1 Results on Pascal VOC Dataset

The Pascal VOC2007 has 9963 images of 20 categories with each image associated with one or more category labels and several describing words. We choose images with single category label for this evaluation, resulting in 5649 images (2808 images for the training set and 2841 images for the test set). Similar to the setting in LCFS and LGCFL methods, we discard the words that appear in less than 3 images, resulting in 399 unique words. For each sample, its image is represented by the 512-dimensional GIST feature extracted in 4×4 blocks, and its text as a 399-dimensional index vector of selected words.

In this subsection, we take the PLS, CCA, GMMFA, GMLDA, LCFS, LGCFL and DLCLA as counterparts. report the results for *Img2Text* and *Text2Img* tasks. To to make fair comparison, for PLS, CCA, GMMFA, GMLDA methods, we perform Principal Component Analysis (PCA) on the original image and text features with 95 % information energy preserved, to remove redundant features; while for LCFS, LGCFL and DLCLA methods, we preserve the original features of image and text modalities.

For the input multi-modal data used in the proposed method, we consider two schemes: (1) linear scheme that use the original multi-modal features; (2) nonlinear scheme that applys RBF kernel mapping to the original multi-modal features. The consideration of kernel mapping is to better capture underlying nonlinear structure of the original multi-modal features.

The MAP scores of the cross-modal retrieval results are shown in Table 2. We can see that our proposed method (linear case and nonlinear) significantly outperforms the previous methods. It shows that our coupled dictionary learning algorithm has the advantage of outputting sparse representation that preserves the relationship among different modalities. In addition, we observe that the nonlinear scheme outperforms the linear scheme. This indicates that applying a proper nonlinear mapping to the original features may significantly improve the retrieval performance. In our case, RBF kernel mapping boosts the performance of Img2Text by about 3 %.

Figure 3 shows the precision-scope curves of our nonlinear scheme against previous approaches for both Img2Text and Text2Img retrieval tasks. The scope (top K retrieved samples) varies from 50 to 1000. For Img2Text retrieval, we observe that our nonlinear scheme constantly outperforms previous approaches. For Text2Img retrieval task, our nonlinear scheme runs tight to LGCFL and constantly outperforms other approaches; however, we outperform LGCFL significantly on certain K values, showing the superiority of our approach.

Table 2 MAP scores of different methods on Pascal VOC2007 dataset

Method	Img2Text	Text2Img	Average
PCA+PLS	0.276	0.199	0.238
PCA+CCA	0.265	0.221	0.243
PCA+GMMFA	0.309	0.231	0.270
PCA+GMLDA	0.242	0.204	0.223
LCFS	0.344	0.267	0.306
LGCFL	0.401	0.322	0.362
DLCLA	0.382	0.317	0.350
Proposed (linear)	0.384	0.325	0.354
Proposed (nonlinear)	0.416	0.337	0.366

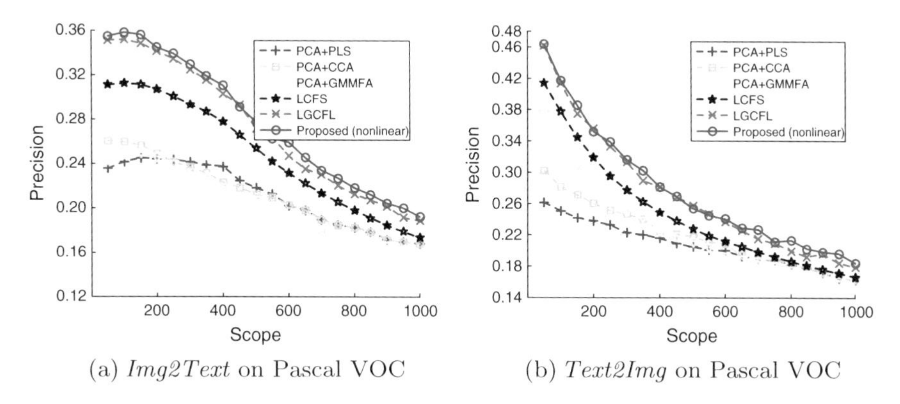

Fig. 3 Precision-scope curves of different methods on the Pascal VOC dataset

4.1.2 Results on Wiki Dataset

The Wiki dataset is collected from the articles on Wikipedia and contains 2866 image-text pairs from ten different classes. The publicly available features from the website[1] are used in the experiment. For each sample in the dataset, it belongs to one of the ten classes, and its image is represented by 128-dimensional SIFT histogram and its text is represented by 10-dimensional vector of topic proportions generated by the LDA [47] model. Similar to the setting in LGCFL method, we randomly selected one hundred samples in each class for training and the rest for test, resulting a split of 1000 ad 1866 samples for training and test sets. Due to the low dimensionality of the original image and text features, we directly use theses features without preprocessing of PCA for all the compared methods.

Table 3 shows the MAP scores of different approaches on the Wiki dataset. On average, our proposed schemes achieves higher MAP scores than the other methods

[1]http://www.svcl.ucsd.edu/projects/crossmodal.

Table 3 MAP scores of different methods on Wiki dataset

Method	Img2Text	Text2Img	Average
PLS	0.245	0.189	0.217
CCA	0.235	0.181	0.208
GMMFA	0.256	0.195	0.226
GMLDA	0.255	0.194	0.225
LCFS	0.268	0.215	0.242
LGCFL	0.279	0.217	0.248
DLCLA	0.264	0.224	0.244
Proposed (linear)	0.264	0.212	0.238
Proposed (nonlinear)	0.269	0.225	0.247

but performs worse than some methods on *Img2Text* task. As discussed in [42], it is challenging to improve on this dataset due to the low dimensionality of image and text features.

Figure 4 further shows the precision-scope curves of our nonlinear scheme against previous approaches for both Img2Text and Text2Img retrieval tasks. The scope (top K retrieved samples) varies from 50 to 1000. We can observe that for Img2Text retrieval task, our nonlinear scheme runs tight to LCFS and LGCFL and sometimes underperforms one of them. On Text2Img retrieval task, we observe that our nonlinear scheme outperforms all other approaches except for LGCFL—we run tight to LGCFL and even sometimes underperforms it. However, for $K < 200$, our approach still outperforms LGCFL by a significant margin.

Nevertheless, our proposed schemes consistently outperform previous methods especially on *Text2Img* task. To further understand the reason, we change the dictionary size of the coupled dictionary learning step in the nonlinear scheme and compare the performance. Figure 5 shows the MAP scores with different dictionary size on the two datasets. We can see that larger dictionary size generally has better capability for sparse representation and the best MAP scores on the two datasets are achieved with dictionary size 300 and 210, respectively. Therefore, the limitation of low dimensional (10-dim) text features (see Table 1) of the Wiki dataset can be tackled by our coupled dictionary learning procedure where we use more efficient high dimensional (210-dim) sparse features, benefiting the coupled feature selection in the feature mapping procedure. However, the decay on *Img2Text* indicates that the sparse representations of different modalities may need to be further balanced.

For the *Text2Img* task, Fig. 6 shows three examples of text queries and the top five images retrieved by the proposed method (nonlinear case). It can be observed that our method finds the closet matches of the image modality at the semantic level for both text queries. And the retrieved images are all belonging to the same label of the text queries, i.e., “music,” “warfare,” and “geography,” respectively.

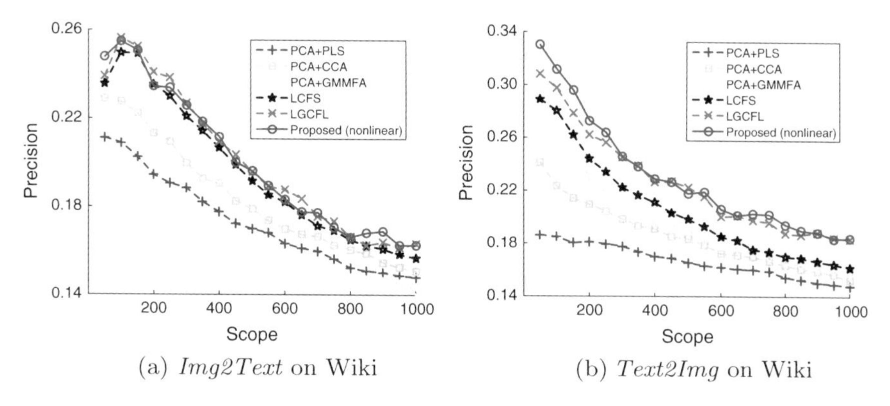

Fig. 4 Precision-scope curves of different methods on the Wiki dataset

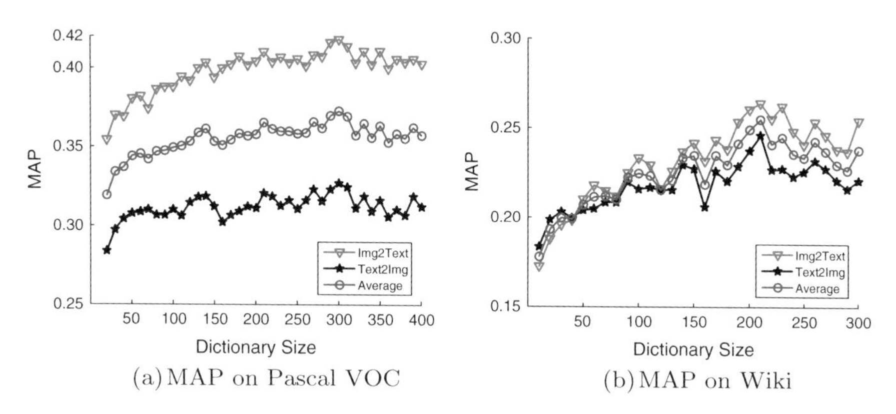

Fig. 5 The MAP with different dictionary size of the proposed method (nonlinear case) on the Pascal VOC dataset (*left*) and Wiki dataset (*right*)

Fig. 6 Typical examples of the *Text2Img* task obtained by our proposed method on Wiki dataset. In each example, the text query and its corresponding image are shown at the *left*, and the *top* five images retrieved are listed in the following columns

4.1.3 Results on MIRFlickr-25K Dataset

MIRFlickr-25K is a real-world dataset that originally consists of 25000 samples collected from Flickr website, each being an image with its associated tags. Besides, each sample is manually annotated with at least one of 38 labels. For the feature representation, we use the features extracted in [5]. Specifically, for each sample, its image is represented by a 3,857-dimensional feature vector by directly concatenating SIFT, Gist and MPEG-7 descriptors provided in [46]; its text input is represented using an indicator vector of the selected 2,000 most frequent tags.

To make comprehensive comparison for the *Img2Label* task, except for the LCFS, LGCFL and DLCLA that are developed for cross-modal retrieval problem, we also take into account several state-of-the-art image annotation methods, which can also be evaluated for the *Img2Label* task. These methods include: (1) JEC [14], Tagprop [13] and Fasttag [48], which only use image modality for learning model; (2) Multi-kernel SVMs (KSVM) [10], which applies different kernel functions to train SVM classifiers for image and text modalities; (3) Kernel CCA (KCCA) [12], which is also a common subspace learning based method. Since KCCA does not directly map image modal data into the label space, we use it for nearest neighbor selection then combine it with the nearest neighbor based tag propagation scheme Tagprop (this has been reported with promising result in [12]). For our proposed method, here we evaluate the linear case due to the high dimensional multi-modal features, and empirically set dictionary size as 350 for coupled dictionary learning.

We report the overall performance on *Img2Label* task in Table 4. We observe that: (1) using an additional text modality improves the accuracy of *Img2Label* task than only using the image modality; (2) our proposed method outperforms the subspace learning method LCFS, indicating that the coupled dictionary learning procedure in our proposed framework is efficient to handle the diversity of different modalities and the learned sparse representations is more powerful than the original features for

Table 4 Overall comparison of three task on the MIRFlickr-25K dataset

Task	Img2Text	Text2Img	Img2Label	
Method	MAP	MAP	P	R
JEC	–	–	0.329	0.173 –
TagProp	–	–	0.452	0.302
Fasttag	–	–	0.450	0.364
KSVM	–	–	0.516	0.366
KCCA+Tagprop	0.579	0.589	0.547	0.354
LCFS	0.533	0.546	0.325	0.312
LGCFL	0.596	0.599	0.523	0.354
DLCLA	0.573	0.587	0.554	0.326
DBM	0.600	0.607	0.581	0.365
Proposed	0.608	0.617	0.576	0.374

Table 5 Comparison of training and test time (in seconds)

	LCFS	LGCFL	DLCLA	Proposed
Training	315.7	375.2	523.4	467.2
Test	0.8	1.5	3.5	2.6

subspace learning; (3) our proposed method achieves the highest performance and generally outperforms the state-of-the-art method KCCA+Tagprop, showing that the learned sparse representations are powerful for subspace learning and coupled feature selection is crucial in enhancing the relationships across different modalities.

In Table 5, we report the processing time of the training and the test stages of the proposed method (linear case) performing on a desktop machine which has 8-core 3.4 GHz CPUs with 32 GB RAM.

From Table 5, we can see that DLCLA and the proposed methods need more time for training and test time, due to the high cost of dictionary learning in each iteration. Both the LCFS and LGCFL have closed-form solutions for optimization and are transductive for testing, hence they require less computing time. In practice, it is possible to use some sophisticated toolbox with multi-core implementation (e.g., SPAMS[2]) to accelerate the training time of the proposed method. Furthermore, caching scheme can be applied to boost the effectiveness of test (retrieval) procedure.

4.2 *Partially Paired Setting*

In this section, we use MIRFlickr-1 M [49] to evaluate the partially paired setting which is a large-scale web image datasets that originally collected for tackling the image annotation problem. Each image in this dataset is associated with corresponding text or document. Specifically, for the MIRFlickr-1M dataset, among the one million images, a subset of 25 K images have been manually annotated with 38 class labels, which is exactly the MIRFlickr-25 K dataset we have used in the *fully paired* setting. The remaining subset of 975 K images have no class labels. Similar to the setting in MIRFlickr-25 K dataset, we represent each image as a 3,857-dimensional concatenated multiple feature, and each text as a 2,000-dimensional binary tagging vector w.r.t the top 2,000 most frequent tags. However, there are 128,501 images without text, since none of the top-frequent tags occurs in the text. Therefore, these images are considered as unpaired samples.

We compare the proposed method with two recently published works LGCFL [24] and DBM [5] that can also handle partially paired setting. DBM utilizes deep neural network with multi-layer Bolzmann machine to learn joint space of the image and text modalities for cross-modal retrieval. Similar to the settings in MIRFlickr-25 K datasets, we evaluate the linear case of the proposed method and empirically

[2]http://spams-devel.gforge.inria.fr/.

set dictionary size as 350 for coupled dictionary learning for these two large-scale datasets. The evaluation is performed for *Img2Text*, *Text2Img* and *Img2Label* tasks.

4.2.1 Results on MIRFlickr-1 M Dataset

We first evaluate the proposed model on MIRFlickr-1 M dataset with the training samples containing both paired samples and unpaired samples. We consider the paired samples with class labels in the 25 K subset as the start point, and gradually add the remaining unpaired samples for training until all the unpaired samples are covered. We report the evaluation results on the test set of the 25 K subset. Specifically, for the LGCFL method we first use the model trained on the 25 K subset to predict class labels for the unpaired samples, and then conduct experiment on the entire dataset of 1 million samples with partially paired setting.

To investigate the effect of utilizing the additional unpaired sampled in the training stage on the retrieval performance, we take different ratio (from 10 to 100 %) of unpaired samples into the training procedure, and evaluate the retrieval result on different tasks. Figure 7a shows the average MAP score of the *Img2Text* and *Text2Img* tasks of the cross-modal retrieval task; and Fig. 7b illustrates the overall F1 score ($F1 = 2\frac{P\times R}{P+R}$) of the *Img2Label* task. We can observe that for all three methods further improvements can be achieved by using more additional data. In particular, for the proposed model, the performance consistently increases when more samples are added. The reason is that our semi-supervised learning framework can incorporate the unimodal data of unpaired samples in the coupled dictionary learning step, which leads to more robust sparse representations for both modalities. The proposed model performs better than LGCFL and is more robust to tackle heterogeneous feature representations of different modalities, whereas LGCFL is elaborately designed for blob based image features and requires class label information even for unpaired samples. Moreover, the proposed model obtains comparable results with deep learning based

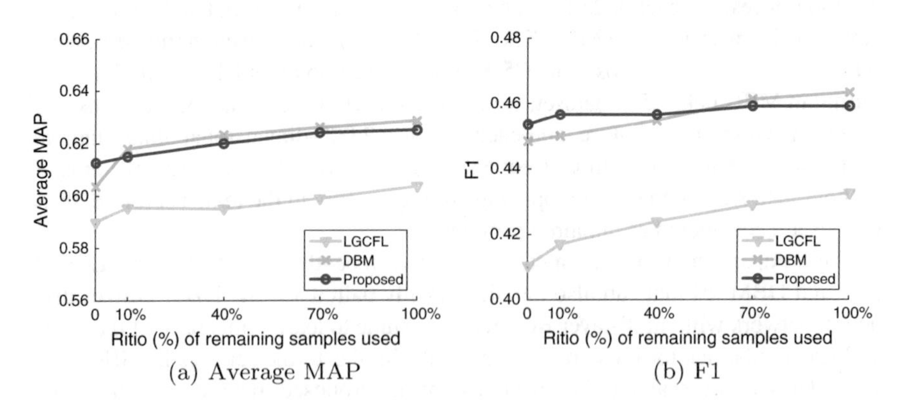

(a) Average MAP (b) F1

Fig. 7 Comparison of using different ratio of unpaired samples for the retrieval performance

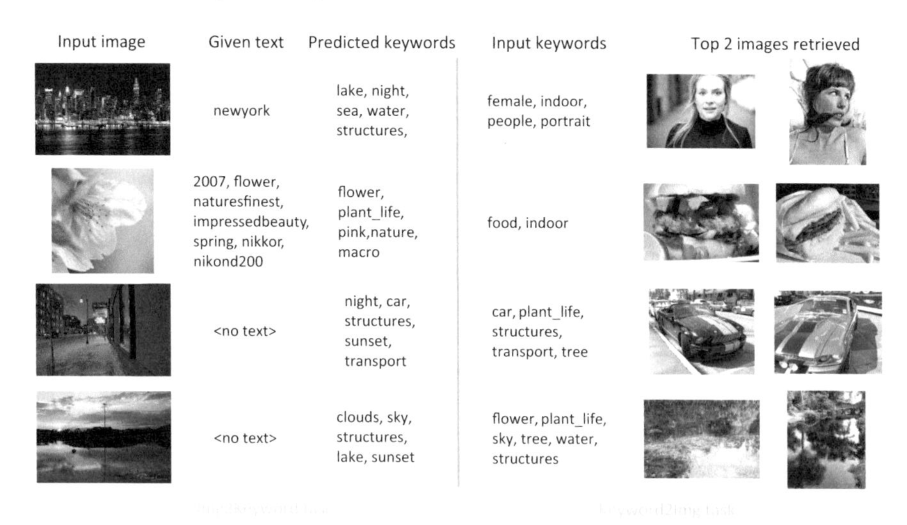

Fig. 8 Examples of the *Img2Label* and *Label2Img* tasks obtained by the proposed advanced scheme on MIRFlickr-1M dataset

Table 6 Overall results on entire collection of 1 million samples under partially paired setting

Task	Cross-modal retrieval		Img2Label	
Method	Img2Text	Text2Img	P	R
LGCFL	0.598	0.609	0.561	0.352
DBM	0.622	0.635	0.591	0.381
Proposed	0.619	0.631	0.582	0.389

DBM approach. Compared with the sophisticated deep architecture in DBM, the proposed model is simpler and more promising to cope with large scale data and to obtain competitive performance under partially paired setting. Figure 8 shows some examples of the retrieval tasks. In each case, the query image or text is shown at the left, and the retrieved texts or images are listed at the following columns. It can be observed that the proposed method is able to find good matches of one modality given a query of another modality, or predict proper words that well describe the visual content of test image (Table 6).

5 Conclusion

In this work, we propose a novel model solving the practical cross-modal retrieval problem. To utilize class label information while simultaneously handling both paired and unpaired samples, we propose a joint learning framework that employs semi-supervised coupled dictionary learning in conjunction with coupled feature mapping schemes. The dictionary learning step generates homogeneously sparse representations for different modalities while the coupled feature mapping step projects the previously generated sparse representations into the class label space for cross-modal matching. We also extend the proposed method to nonlinear case by applying kernel mapping to the original multi-modal features. This scheme has shown its superiority of capturing underlying nonlinear structure of the original multi-modal features and its benefit to the retrieval performance. The proposed method was initially evaluated on three benchmark datasets with fully paired setting and further validated on two large-scale datasets with partially paired setting. Experimental results has verified the effectiveness of the proposed approach on both cross-modal retrieval tasks and image annotation task, which is competitive to several related state-of-the-art methods.

References

1. L. Chen, D. Xu, I.W. Tsang, X. Li, Spectral embedded hashing for scalable image retrieval. IEEE Trans. Cybern. **44**(7), 1180–1190 (2014)
2. F. Shen, C. Shen, Q. Shi, A. Van Den Hengel, Z. Tang, Inductive hashing on manifolds, in *IEEE Conference on Computer Vision and Pattern Recognition* (2013)
3. J. Song, Y. Yang, X. Li, Z. Huang, Y. Yang, Robust hashing with local models for approximate similarity search. IEEE Trans. Cybern. **44**(7), 1225–1236 (2014)
4. Y. Yang, Z.-J. Zha, Y. Gao, X. Zhu, T.-S. Chua, Exploiting web images for semantic video indexing via robust sample-specific loss. IEEE Trans. Multimedia **16**, 1677–1689 (2014)
5. N. Srivastava, R. Salakhutdinov, Multimodal learning with deep boltzmann machines. JMLR **15**, 2949–2980 (2014)
6. Y. Yang, Y. Yang, H.T. Shen, Effective transfer tagging from image to video. ACM Trans. Multimedia Comput. Commun. Appl. **9**(2), 1–20 (2013)
7. Y. Yang, Z.-J. Zha, Y. Gao, X. Zhu, T.-S. Chua, Exploiting web images for semantic video indexing via robust sample-specific loss. TMM **17**(2), 246–256 (2015)
8. Y. Zhen, Y. Gao, D. Yeung, H. Zha, X. Li, Spectral multimodal hashing and its application to multimedia retrieval. IEEE Trans. Cybern. **46**(1), 27–38 (2016)
9. V. Ordonez, G. Kulkarni, T.L. Berg, Im2text: describing images using 1 million captioned photographs. NIPS (2011)
10. J. Verbeek, M. Guillaumin, T. Mensink, C. Schmid, Image annotation with tagprop on the mirflickr set, in *ACM MIR*, MIR '10, pp. 537–546 (2010)
11. Y. Verma, C. Jawahar, Im2text and text2im: associating images and texts for cross-modal retrieval, in *BMVC* (2014)
12. L. Ballan, T. Uricchio, L. Seidenari, A. Bimbo, A cross media model for automatic image annotation. ICMR (2014)
13. M. Guillaumin, T. Mensink, J. Verbeek, C. Schmid, Tagprop: discriminative metric learning in nearest neighbor models for image auto-annotation. CVPR 309–316 (2009)
14. A. Makadia, V. Pavlovic, S. Kumar, A new baseline for image annotation, in *ECCV* (2008)
15. D.R. Hardoon, S.R. Szedmak, J.R. Shawe-taylor, Canonical correlation analysis: an overview with application to learning methods. Neural Comput. **16**(12), 2639–2664 (2004b)
16. Y. Gong, Q. Ke, M. Isard, S. Lazebnik, A multi-view embedding space for modeling internet images, tags, and their semantics. IJCV **106**, 210–233 (2014)
17. N. Rasiwasia, J. Costa Pereira, E. Coviello, G. Doyle, G. Lanckriet, R. Levy, N. Vasconcelos, A new approach to cross-modal multimedia retrieval, in *ACM MM* (2010)

18. H. Zhang, Y. Zhuang, F. Wu, Cross-modal correlation learning for clustering on image-audio dataset, in *Proceedings of the 15th ACM International Conference on Multimedia*, pp. 273–276 (2007)
19. D. Li, N. Dimitrova, M. Li, I.K. Sethi, Multimedia content processing through cross-modal association, in *Proceedings of the Eleventh ACM International Conference on Multimedia*, pp. 604–611 (2003)
20. S. Deerwester, S.T. Dumais, G.W. Furnas, T.K. Landauer, R. Harshman, Indexing by latent semantic analysis. J. Amer. Soc. Inf. Sci. **41**(6), 391–407 (1990)
21. R. Rosipal, N. Krämer, Overview and recent advances in partial least squares, in *Proceedings of SLSFS*, pp. 34–51 (2006)
22. A. Sharma, D.W. Jacobs, Bypassing synthesis: pls for face recognition with pose, low-resolution and sketch, in *Proceedings of CVPR*, pp. 593–600 (2011)
23. J.B. Tenenbaum, W.T. Freeman, Separating style and content with bilinear models. Neural Comput. **12**(6), 1247–1283 (2000)
24. C. Kang, S. Xiang, S. Liao, C. Xu, C. Pan, Learning consistent feature representation for cross-modal multimedia retrieval. IEEE Trans. Multimedia **17**(3), 370–381 (2015)
25. X. Gao, N. Wang, D. Tao, X. Li, Face sketch-photo synthesis and retrieval using sparse representation. IEEE Trans. Circuits Syst. Video Technol. **22**(8), 1213–1226 (2012)
26. Z. Cui, W. Li, D. Xu, S. Shan, X. Chen, X. Li, Flowing on riemannian manifold: domain adaptation by shifting covariance. IEEE Trans. Cybern. **44**(12), 2264–2273 (2014)
27. D. Huang, Y. Wang, Coupled dictionary and feature space learning with applications to cross-domain image synthesis and recognition, in *ICCV*, pp. 2496–2503 (2013)
28. S. Shekhar, V.M. Patel, H.V. Nguyen, R. Chellappa, Coupled projections for adaptation of dictionaries. IEEE Trans. Image Process. **24**(10), 2941–2954 (2015)
29. S. Wang, L. Zhang, Y. Liang, Q. Pan, Semi-coupled dictionary learning with applications to image super-resolution and photo-sketch synthesis. CVPR 2216–2223 (2012)
30. J. Wang, S. Kumar, S.-F. Chang, Semi-supervised hashing for scalable image retrieval. CVPR 3424–3431 (2010)
31. A. Sharma, A. Kumar, H. Daume, D.W. Jacobs, Generalized multiview analysis: a discriminative latent space. CVPR 2160–2167 (2012)
32. M. Kan, S. Shan, H. Zhang, S. Lao, X. Chen, Multi-view discriminant analysis, in *Computer Vision ECCV 2012*, vol 7572, pp. 808–821 (2012)
33. C. Deng, X. Tang, J. Yan, W. Liu, X. Gao, Discriminative dictionary learning with common label alignment for cross-modal retrieval. IEEE Trans. Multimedia **18**(2), 208–218 (2016)
34. X. Xu, A. Shimada, R. Taniguchi, L. He, Coupled dictionary learning and feature mapping for cross-modal retrieval, in *IEEE International Conference on Multimedia and Expo, ICME*, pp. 1–6 (2015)
35. D. Hardoon, S. Szedmak, J. Shawe-taylor, Canonical correlation analysis: An overview with application to learning methods. Neural Comput. **16**, 2639–2664 (2004a)
36. H. Hotelling, Relations between two sets of variates. Biometrika (1936)
37. J.C. Pereira, E. Coviello, G. Doyle, N. Rasiwasia, G.R.G. Lanckriet, R. Levy, N. Vasconcelos, On the role of correlation and abstraction in cross-modal multimedia retrieval. IEEE Trans. Pattern Anal. Mach. Intell. **36**(3), 521–535 (2014)
38. S. Hwang, K. Grauman, Learning the relative importance of objects from tagged images for retrieval and cross-modal search. IJCV **100**, 134–153 (2012)
39. L. Zhang, Y. Zhao, Z. Zhu, S. Wei, X. Wu, Mining semantically consistent patterns for cross-view data. IEEE Trans. Knowl. Data Eng. **26**(11), 2745–2758 (2014)
40. Y. Chen, L. Wang, W. Wang, Z. Zhang, Continuum regression for cross-modal multimedia retrieval. ICIP (2012)
41. L. Sun, S. Ji, J. Ye, A least squares formulation for canonical correlation analysis. ICML 1024–1031 (2008)
42. K. Wang, R. He, W. Wang, L. Wang, T. Tan, Learning coupled feature spaces for cross-modal matching, in *ICCV*, pp. 2088–2095 (2013)

43. J. Ngiam, A. Khosla, M. Kim, J. Nam, H. Lee, A.Y. Ng, Multimodal deep learning. ICML 689–696 (2011)
44. N. Srivastava, R. Salakhutdinov, Multimodal learning with deep boltzmann machines, in *NIPS*, pp. 1–9 (2012)
45. Y. Peng, X. Zhai, Y. Zhao, X. Huang, Semi-supervised cross-media feature learning with unified patch graph regularization. IEEE Trans. Circuits Syst. Video Technol. **26**(3), 583–596 (2016)
46. M.J. Huiskes, M.S. Lew, The mir flickr retrieval evaluation, in *ACM MIR*, pp. 39–43 (2008)
47. D.M. Blei, A.Y. Ng, M.I. Jordan, Latent dirichlet allocation. J. Mach. Learn. Res. **3**, 993–1022 (2003)
48. M. Chen, A. Zheng, K. Weinberger, Fast image tagging. ICML 1274–1282 (2013)
49. M.J. Huiskes, B. Thomee, M.S. Lew, New trends and ideas in visual concept detection: The mir flickr retrieval evaluation initiative, in *ACM MIR*, pp. 527–536 (2010)

Light Field Vision for Artificial Intelligence

Yichao Xu and Miu-ling Lam

Abstract Light field camera has been available on the market, and the its capability of capturing both spatial and angular information makes it more powerful for solving computer vision problems. A newly developed *Light Field Vision* technique shows a big advantage over conventional computer vision techniques. We review the recent progress in *Light Field Vision*.

Keywords Light field · Image acquisition · Computer vision

1 Introduction

Computer vision techniques, which are inspired by theories and observations of visual perception, have been developed rapidly and applied for many kinds of artificial intelligent (AI) applications since it appeared in 1966 [1]. The computer vision systems acquire the image, video and multi-dimensional data from the vision sensors, and they can apply the theories and models to solve various problems. Typical computer vision problems include object recognition, scene understanding, video tracking, motion estimation and so on. The solutions to these problems are very useful for the AI systems, and computer vision usually plays an important role in the intelligent robotics.

Nowadays, the intelligent systems are not far from our daily life, and they can make our life better. For example, when we use a smart phone to take a photo of a monument as shown in Fig. 1a, the computer vision system can tell us the related knowledge of the photo, such as what is the monument for and who made this monument, and then we can know more about what we see. The driverless car will be true with the help of computer vision system as shown in Fig. 1b.

Y. Xu (✉) · M.-l. Lam
School of Creative Media, City University of Hong Kong,
Kowloon Tong, Hong Kong
e-mail: yichaoxu@cityu.edu.hk

H. Lu and Y. Li (eds.), *Artificial Intelligence and Computer Vision*,
Studies in Computational Intelligence 672, DOI 10.1007/978-3-319-46245-5_11

(a) Smart phone [2]. (b) Driverless car [3].

Fig. 1 Two examples of vision-based AI applications. **a** Smart phone [2]. **b** Driverless car [3]

A recently developed vision sensor, which is called light field camera, can capture richer information from the 3-Dimensional (3D) world than conventional cameras. The light field camera can record light rays from every direction through every point in the 3D world. Therefore, we can deal with more challenging AI applications with the data captured by the light field camera.

1.1 Light Field Vision

As Adelson and Bergen pointed out visual information available to an observer at any point in space and time [4]. Actually, objects can emit or reflect light rays, and we call all the light rays in the space ***light field***. The ***light field*** includes all the visual information in the space.

The visual data is acquired by various vision sensors, and the data can be taken in many kinds of forms. The charge-coupled device (CCD) has been widely used in digital image sensing, because we can use the CCD image sensor to acquire high-quality images and video sequences with low cost. CCD image sensors are easy to use since there are many user-friendly hardware and software available. However, conventional CCD image sensor can only capture sub light field space as shown in Fig. 2. On the contrary, an active sensor requires an external source of power to send out a signal to be bounced off a target, and detects the reflected signals. Such kind of sensors include laser scanner, radar, sonar and time-of-flight camera etc. These active sensors can acquire more information, like the depth of target object, but these devices usually require extra power supply and dedicated software.

However, the light field camera can capture both visual and depth information with passive sensor, and it has been used for a variety of different visualization applications, such as generating free-view images, 3D graphics, and digital refocusing in the computer graphics community. Because the light field image records richer

Fig. 2 Each viewpoint can only capture sub light field space

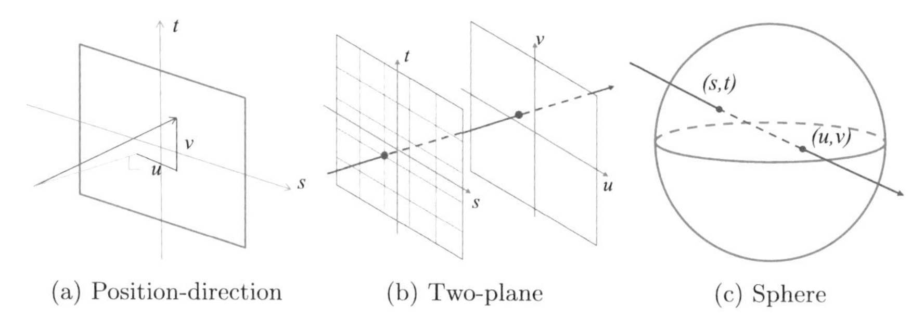

Fig. 3 Different parameterizations of the light field. **a** Position-direction. **b** Two-plane. **c** Spherical

information than that captured by conventional cameras, and the equipment is easy to get nowadays, we believe that light field is useful in computer vision applications, and such kind of applications are becoming popular.

The technique that utilizes light field data to solve computer vision problems is called ***light field vision***.

The light rays in the 3D world can be parameterized in the 4D coordinates, and each ray is represented by $L(s, t, u, v)$. There are several different ways to parameterize the light rays, such as position-direction [5], two-plane [6], and spherical [7] parameterizations (see Fig. 3).

Since the data captured by light field cameras has richer information than that captured by conventional cameras, light field cameras are becoming popular in computer vision applications. The comparison of regular computer vision and light field vision are shown in Fig. 4. The regular computer vision applications are based on the images captured by single viewpoint camera as shown in Fig. 4a. The actual 3D

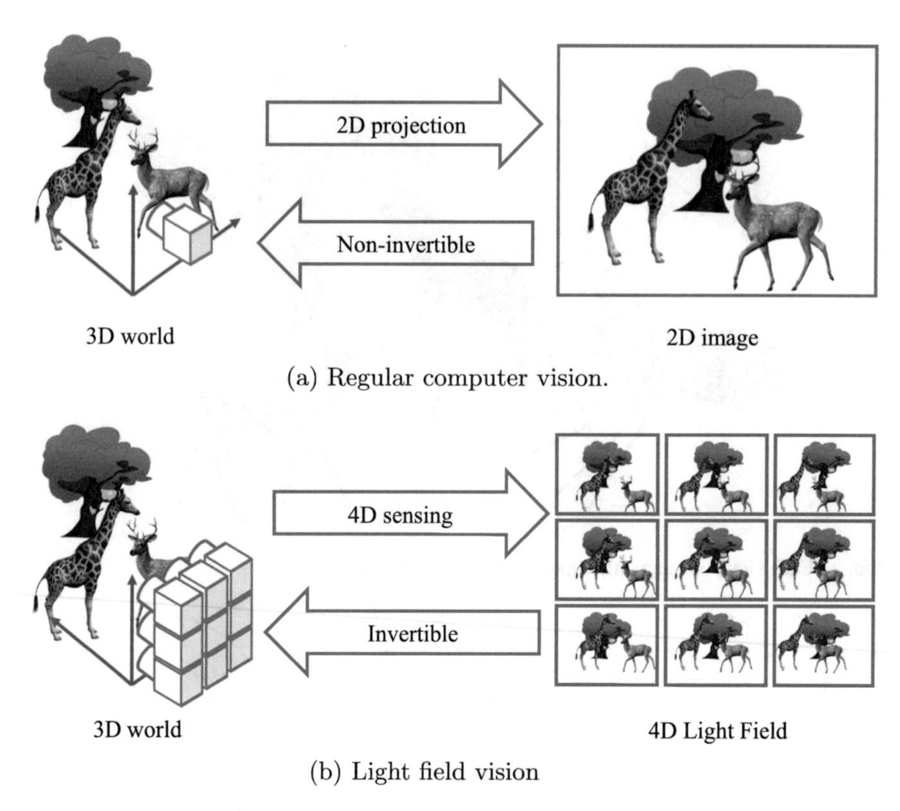

Fig. 4 Regular computer vision and light field vision. **a** Regular computer vision. **b** Light field vision

scene is projected to a 2D image. The depth information of the light rays disappear after the projection. Consequently, we cannot know how far is the object from a single image, and it is difficult to recognize objects and scenes in the real 3D world from the image. If we use a light field camera to capture the data as shown in Fig. 4b. The light field image maintain the 2D positional information, and 2D directional information of light rays from the 3D scene. The redundant information makes it easier to understand the 3D world.

1.2 Pipeline of Light Field Vision Applications

There are four basic stages when we apply the light field vision to AI applications as shown in Fig. 5.

Firstly, the light field data should be captured. People use light field camera to capture 4D light rays. In the early days, light field was obtained by a camera on a

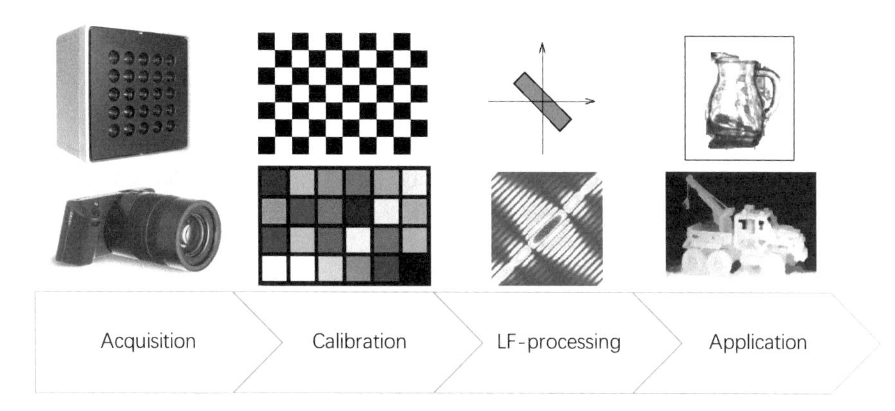

Fig. 5 Pipeline of light field vision applications

gantry. The large camera array systems was developed in the beginning of 21st century, e.g. Stanford multi-camera array. These light field camera systems were usually huge and quite expensive. Fortunately, recent light field cameras are becoming inexpensive and compact, such as Profusion25 [8], Lytro [9], and Raytrix [10]. We will discuss more about the light field acquisition systems in Sect. 2.

Camera calibration is an essential step in light field vision applications, no matter using which kind of camera to acquire light field images. Since light field camera can capture multiple viewpoints of the scene, not only geometric relationship but also the photometric consistency should be considered in this calibration stage. After geometric calibration, the relationship between the captured light rays become known, while the photometric calibration can make the color information consistency for a same point. We will discuss the calibration methods for light field cameras in Sect. 3.

After camera calibration, we can do some light field processing with the known camera parameters. Similar as conventional image processing, 4D light field processing can be performed in spatial domain as well as in the frequency domain. We can transform the captured light rays into a certain space which is helpful for the the applications. The light field processing includes light ray resampling, filtering in spatial and frequency domain. For the light field video, we can also perform video stabilization in the temporal domain. Details of light field processing will be discussed in Sect. 4.

Researchers have used light field cameras for computer vision applications in the recent years, such as surveillance [11], consistent depth estimation [12], salience detection [13] and transparent object segmentation [14]. And these applications show that light field vision has better performance than previous computer vision approaches. More examples can be found in Sect. 5 where the authors review the recent work of light field vision for AI applications.

2 Light Field Acquisition

As mentioned in in previous section, the first stage of light field vision applications is to acquire light field data. There are various types of systems can obtain light field data. Different applications require different light field data. The light field acquisition systems is reviewed in this section.

2.1 Gantry Camera

A simple way to acquire the light field is to put a camera on the moving gantry as shown in Fig. 6. Researchers at Stanford University first built a gantry (Fig. 6a) for light field rendering [6], and the specifications of their gantry are available on the website of Cyberware [15]. Researchers from Cornell University then built a gantry (Fig. 6b) that improves the mounting arrangement at the ends of the arms [16]. It makes more flexibility in the lamp and camera that can be attached to these arms. Researchers from University of Virginia have also built a gantry (Fig. 6c) with similar design, but the light and camera are coaxially mounted to each of the arms [17].

The cost of a spherical gantry is very high, and it is not worthy to build such expensive equipment only for light field acquisition. Researchers found some simple and inexpensive ways to acquire the light field. The Lego Mindstorms gantry can be used to capture a light field (Fig. 7a). We can just move a camera left, right, up, and down on the Lego gantry. And the researchers from MIT graphics group also built a simple vertical XY-table to capture the light field (Fig. 7b).

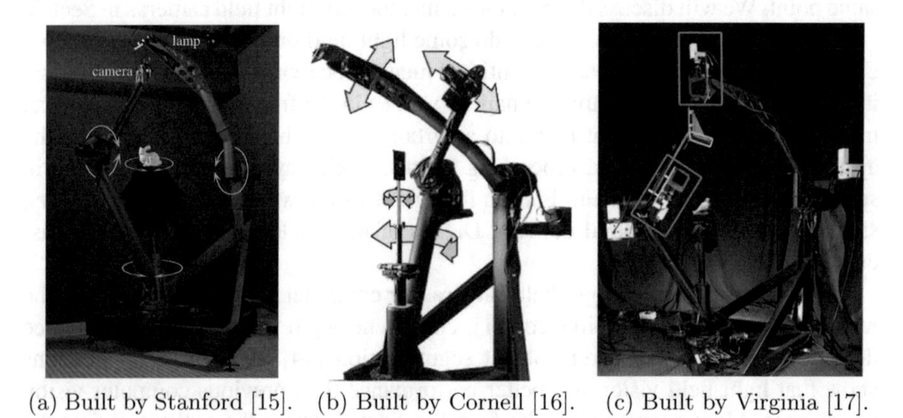

(a) Built by Stanford [15]. (b) Built by Cornell [16]. (c) Built by Virginia [17].

Fig. 6 Spherical gantry cameras for light field acquisition. **a** Built by Stanford [15]. **b** Built by Cornell [16]. **c** Built by Virginia [17]

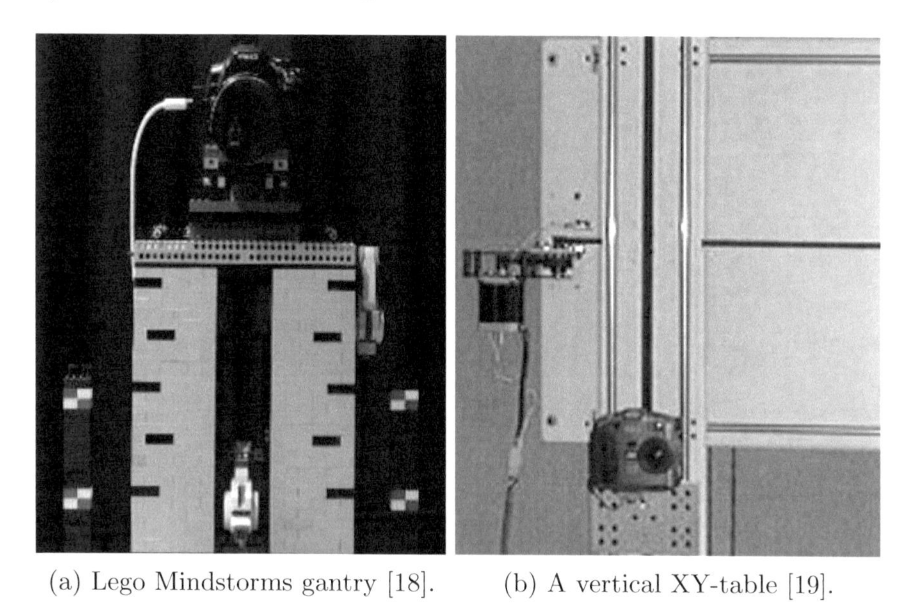

(a) Lego Mindstorms gantry [18]. (b) A vertical XY-table [19].

Fig. 7 Planar gantry cameras for light field acquisition. **a** Lego Mindstorms gantry [18]. **b** A vertical XY-table [19]

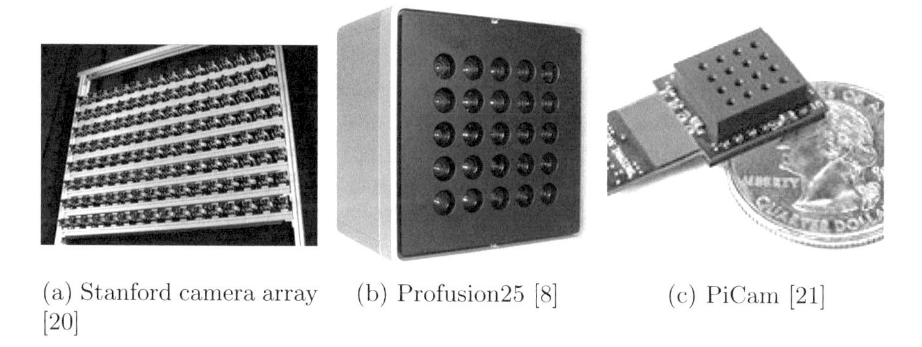

(a) Stanford camera array [20] (b) Profusion25 [8] (c) PiCam [21]

Fig. 8 Camera array systems. **a** Stanford camera array [20]. **b** Profusion25 [8]. **c** PiCam [21]

2.2 *Camera Array*

Moving a single camera on the gantry can only capture the static scenes. In order to capture dynamic scenes, camera array systems (as shown in Fig. 8) have been developed to acquire the light field. Researchers at Stanford University built several large camera array to perform computational photography applications [20]. These camera array systems allow them to capture light field video. In the recent years, camera array systems are becoming more compact. A camera array with 25 viewpoints,

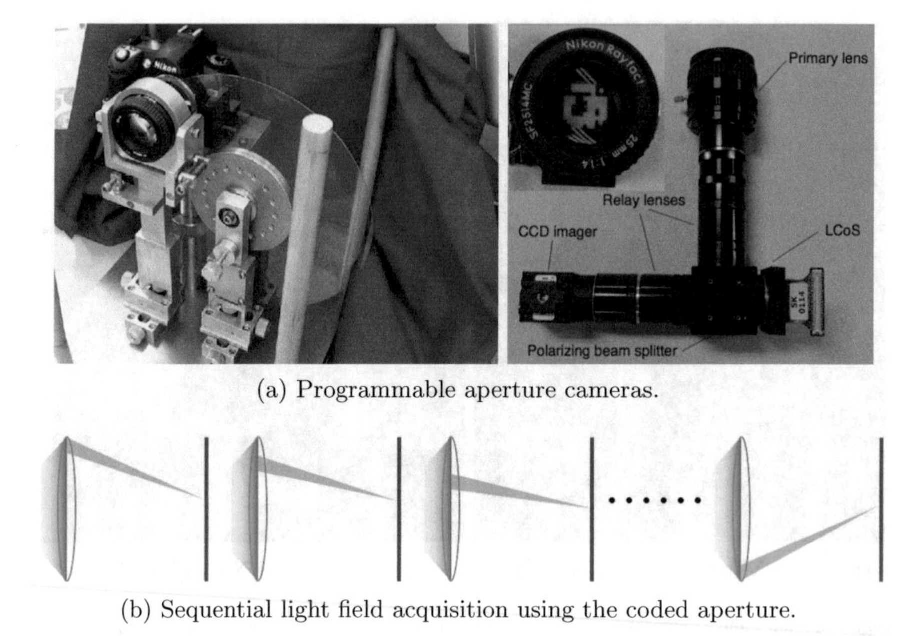

(a) Programmable aperture cameras.

(b) Sequential light field acquisition using the coded aperture.

Fig. 9 Examples of coded aperture cameras and their sequential sampling mode. **a** Programmable aperture cameras. **b** Sequential light field acquisition using the coded aperture

which is called Profusion25, has already been available in the commercial market [8]. It can be easily connected to a Desktop PC or a laptop, which do not require specific control and synchronizing equipment like the large camera array. The latest camera array is even smaller than a coin as shown in Fig. 8c. This camera supports both stills and video, low light capable, and it is small enough to be included in the next generation of mobile devices including smartphones [21].

2.3 Coded Aperture Camera

Light field can also be captured by a camera with a series different aperture shapes. A straightforward way is to blocks all undesirable light rays and leaves a subset of light field to be obtained one by one, as shown in Fig. 9b. This can be simply realized by programmable aperture cameras [22, 23] as shown in Fig. 9a, but the light efficiency is very low because of the small aperture size. In order to overcome this limitation, some well designed aperture shapes is used to capture the images and light field can be recovered by computational methods [24, 25].

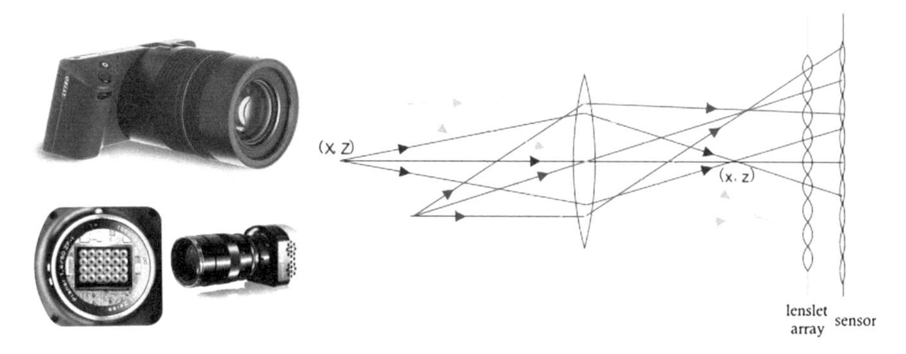

Fig. 10 The lenslet-based light field camera and its projection model

2.4 Lenslet Camera

Similar to the gantry camera, the camera with coded aperture camera cannot capture dynamic scenes as well. However, if we put a lenslet array in front of the image sensor as shown in Fig. 10, the camera can capture the light field with one shot. Ng et al. proposed the first hand-held lenslet camera [26], and this is the prototype of the commercial light field camera Lytro [9]. However, the resolution of first generation lenslet camera is pretty low. The second generation lenslet camera, which is called focused plenoptic camera, has been proposed by Georgiev et al. [5, 27] to increase the image resolution. The focused plenoptic camera is also available on commercial market [10].

3 Light Field Camera Calibration

As introduced in the previous section, light field can be captured by many types of cameras. Some light field acquisition systems are made by researchers themselves, and some of them can be bought from the commercial market. No matter using what kind of camera to acquire light field images, the calibration is an essential step in computer vision applications. Since light field camera can capture multiple viewpoints of the scene, not only geometric relationship but also the photometric consistency should be considered in the calibration stage.

3.1 Geometric Calibration

Camera array calibration Over the past few decades, a great deal of work has been done on camera geometric calibration to acquire camera parameters with high accuracy. There are several camera calibration approaches including the single camera, multi camera, and structure from motion (SfM), which can be directly applied to the camera array, approaches.

Classic camera calibration is performed by observing a 3D reference object with a known Euclidean geometry [28]. This type of approach requires specialized and expensive equipment with an elaborate setup. To overcome these disadvantages, a flexible technique for single camera calibration was proposed by Zhang [29], which requires the camera to observe a planar pattern displayed at a minimum of two different orientations only. The pattern can simply be printed using a laser printer and then attached to a "reasonable" planar surface (e.g., a hard book cover). Either the camera or the planar pattern can then be moved by hand. The specific motion need not be known. Although this technique is very practical and robust for a single camera, it is not suitable for a light field camera. The rigid transformations between any pair of viewpoints, which can be determined using any captured frame, should be invariant irrespective of the frame by which they were computed. Unfortunately, these transformations are inconsistent when each viewpoint is calibrated independently (see Fig. 11). This inconsistency results in inaccurate estimation of the relative translation between the viewpoints, potentially leading to serious problems if used with light field cameras.

Stereo camera is the simplest multi-camera system, and calibration methods utilizing different constraint were proposed for stereo calibration. Horaud et al. [30] proposed a method for recovering camera parameters from rigid motions. This method relies on linear algebraic techniques and requires the epipolar geometry. Malm and Heyden proposed a method [31] which extends Zhang's single camera calibration method, and also utilizing a planar object. Several methods have been developed to deal with multi-camera systems. Vaish et al. [32] proposed a method using a plane plus parallax to calibrate a multi-camera array for light field acquisition. Assuming that the images of the light field were aligned on some reference plane in the world, they were able to measure the parallax of some points in the scene not lying on this reference plane. This method, however, assumes that all cameras lie on a plane parallel to the reference plane, and the projection to the reference plane must be calculated

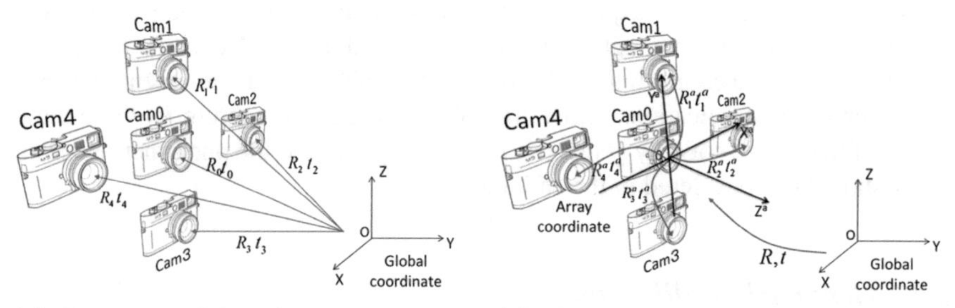

(a) Camera model without rigid constraint (b) Camera model with rigid constraint

Fig. 11 Camera array based light field acquisition system geometry. **a** Camera model without rigid constraint. **b** Camera model with rigid constraint

in advance. Svoboda et al. [33] proposed a method for multi-camera system calibration using point light source. They captured image sequences of the multi-camera while point light source moving in a working volume. The method used the factorization method for solving projective matrices as well as the light source positions from the sequences. Ueshiba et al. [34] proposed a method that uses a planar checkerboard pattern like other methods [29, 35]. They calculate homography matrices between the calibration chart and the images captured by the multi-cameras, then also apply the factorization method to estimating checkerboard chart positions and the projection matrices from the homographies. Xu et al. [35] adopted a camera array model with a rigid constraint between the viewpoints (see Fig. 11b) to calibrate the relative relationship between the viewpoints and other intrinsic and extrinsic parameters of the camera array system.

SfM techniques aim to reconstruct simultaneously the unknown 3D scene structure and camera positions and orientations from a set of feature correspondences. Related methods such as bundle adjustment have made their way into computer vision and are now regarded as the gold standard for performing optimal 3D reconstruction from correspondences [36]. Bundler [37] is a popular tool for SfM. It can also estimate camera parameters from multi-images by bundle adjustment. Bundler was designed for applying either a moving camera or multiple cameras. It has great flexibility in that each viewpoint of the captured image can be freely moving. There are no constraints on camera positions and it independently estimates the multi-camera parameters. As a result of the flexibility, Bundler needs high computational cost while loses calibration accuracy by not using a rigid camera constraint.

Lenslet camera calibration With the availability of commercial lenslet light field camera on the market, it has inspired many vision applications, as well as the research on geometric calibration of lenslet cameras, although there are still only a few.

Dansereau et al. proposed a geometric calibration approach for Lytro cameras [38]. They modeled pixel-to-ray correspondences of the lenslet light field cameras in 3D space, and presented a 4D intrinsic matrix from combining a pinhole model of the lenslet array and thin-lens model of the main lens. Cho et al. also proposed a step-by-step calibration pipeline for Lytro camera calibration [39]. They mainly focus on estimating the rotation angle and the center of the lenslet in the raw light field image. Johannsen et al. presented a geometric calibration method using a dot pattern with a known grid size and a depth distortion correction for Raytrix cameras [40], but they can only estimate the virtual depth range. Zeller et al. also developed a method [41] to calibrate a Raytrix camera similar as [40]. Nevertheless, they did not investigate the distortion of the depth map by the main lens since their method focused on large object distances that the depth map distortion can be neglected. Bok et al. presented a method utilizing raw lenslet images directly [42]. Line features are extracted from raw images and the initial solution of both intrinsic parameters and extrinsic parameters of the lenslet camera projection model is computed by a linear method. The initial solution is then refined via a non-linear optimization. Strobl and Lingenauber developed a method by uncoupling the parameters which can be estimated with conventional 2D calibration methods from the parameters which are specific to the lenslet cameras, like the depth distortion and lenslet parameters [43]. However, this method require known size of the pixel element.

3.2 Photometric Calibration

Photometric calibration is also known as color mapping. It can reproduce the same color values when two or more cameras in the acquisition system. It is very important for the applications of image registration, object tracking and recognition with multiple cameras. Since the lenslet light field camera use a single sensor, the color inconsistency problem mainly caused by the aberration. Here we focus on the color calibration of camera array.

Similar as geometric calibration, a straightforward color calibration way is to use a Macbeth color checker. All the cameras take an image of the Macbeth color checker, and one of these acquired images is considered as the reference image. Joshi used color charts to improve color consistency of large camera array [44]. The gain and offset settings of each camera response are iteratively adjusted on each channel to fit a line through the RGB values recorded the color chart which is put in the static lighting condition. The non-linear error can be corrected by re-mapping the color values with a look-up table regarding to the non-linear sensor response. Inspired by this work, Ilie et al. [45] proposed a two-stage color calibration method for a multi-camera system. Similar as [44], the first stage consists of adjusting configurations of each camera by minimizing color differences between the reference image and images of the color chart acquired by all cameras. When the initial cameras parameters are obtained, same optimization process for all cameras is repeated by comparing with a new reference image calculated as the average of all cameras images computed in the previous step. The second stage uses linear least squares matching, linear color transform and polynomial transform to refine the color consistency between different viewpoints. However, this method requires a constant lighting condition, which cannot be guaranteed in a real practical system.

The methods mentioned above, require to adjust the hardware setting of the cameras, while pure software-based color correction methods are more convenient. Histogram matching is one of the popular methods. Chen et al. [46] proposed an approach based on histogram matching. Two parameters in their approach, namely scaling and offset, can be optimally derived for each YUV channel based on the histograms of the reference viewpoint and other viewpoints which need to be corrected. Similarly, Fecker et al. proposed an extended the histogram matching method that can deal with color correction in both spatial and temporal direction. Lookup tables are calculated with the cumulative histograms of the viewpoints being corrected and the reference. In order to maintain the color consistency in temporal correlation, the authors proposed a time-constant mapping function. They compute the histogram for the whole sequence so that the same correction is used for all temporal frames. Fezza et al. [47] proposed a method using a customized histogram matching, and their method can deal with the occlusion problem. Only common regions across

Fig. 12 An example of color correction result. *Top* images without color correction. *Bottom* images after color correction with the method [47]. The one with a *red frame* is the reference image

viewpoints are taken into consideration when calculating the histogram, thus it can avoid the error from the occlusion regions, which leads to better color correction. The common areas are detected by the scale invariant feature transform (SIFT) [48], followed by the random sample consensus (RANSAC) to remove matching outliers. To keep the temporal correlation, histogram matching is performed on a sliding window, where each color mapping function is defined using a group of images which is similar as [49]. An example of color correction result is shown in Fig. 12.

Recently, a comprehensive review on color correction for multi-view cameras can be found in [50].

4 Light Field Processing

Although traditional 2D image processing algorithms can apply to the light field images one by one viewpoint, the relationship between the viewpoints are not taken into account by those algorithms. In this section, we will focus on the image processing methods specific to 4D light field images.

Digital refocusing Since the light field camera can capture 4D light rays of the scene, it has the capability to re-arrange the obtained light rays after capturing the light field images. We can refocus the image to desired depth and the objects in other depth will be blurred out. The digital refocusing process can be performed in both spatial and frequency domain [26, 51]. The spatial domain implementation integrates the captured light rays onto a certain depth, and the pixels from that depth level will be focused while others will be blurred out. The frequency domain implementation also called *Fourier slice photography*. In the Fourier domain, a photograph formed by a full lens aperture is a 2D slice in the 4D light field. Images focused at different depth levels correspond to the slices at different trajectories in the 4D light field space. An example of digital refocused at different depth levels is shown in Fig. 13.

Digital refocusing technique is helpful in the applications such as surveillance [11, 52] and depth recovery [53].

Super-resolution Light field acquisition must consider the tradeoff between spatial and angular resolution. The resolution of light field images is important for some applications like object detection and classification, in which low resolution images will cause error detection or classification results. However, the lenslet light field camera shares a single sensor to capture the spatial and angular information of the light rays which results in the low spatial resolution of a 2D image. Various super-resolution methods have been developed to overcome the limitation of low spatial resolution [54–58]. An example of light field super-resolution result is shown in Fig. 14.

Bishop et al. [54, 55] developed a variational Bayesian framework to super-resolve the reconstructed viewpoints by fusing multi-view information. Improved demosaicing process for the raw light field images can also super-resolve the reconstructed viewpoints [56, 57]. The analysis and implementation method in spatial domain is describe in [56], while the frequency domain analysis and implementation is given in [57]. Wanner and Goldluecke [58] developed a continuous variational framework for the analysis of 4D light fields. Their method can increase the sampling rate of the 4D light field in spatial as well as angular resolution. Regarding to the angular super-resolution, early in the light filed rendering, Levoy and Hanrahan [6]

Fig. 13 An example of digital refocused at near, middle and distant object

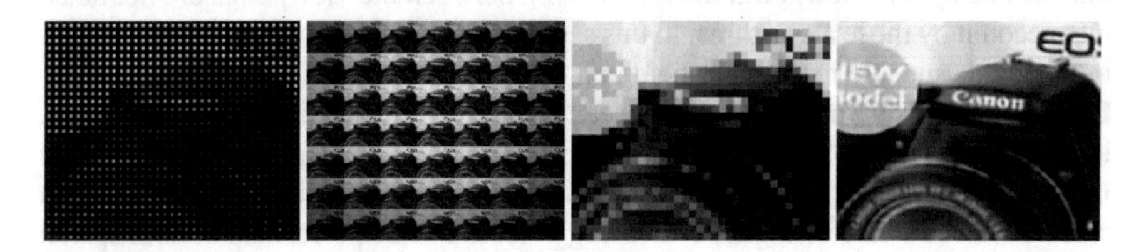

Fig. 14 An example of light field super-resolution result. *Left* Light field image captured with a lenslet light field camera; *Middle left* rearranged light field image shown in multiple viewpoints; *Middle right* central viewpoint extracted from the light field, with one pixel per microlens, as in a traditional rendering [26]; *Right* central view super-resolved with the method [54]

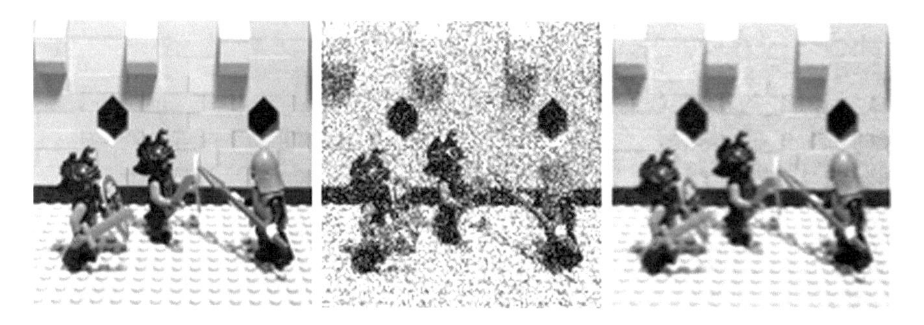

Fig. 15 An example of light field denoising result for Stanford "Lego Knights" light field. *Left* image from original light field; Middle: image with additive white Gaussian noise; *Right* denoising result by the method in [62, 63]

has mentioned that novel viewpoints can be interpolated from the obtained nearby spatial and angular information. Levin and Durand [59] proposed a method that can generate novel viewpoints from 3D focal stack or a sparse set of viewpoints based on a dimensionality gap Gaussian prior. Novel viewpoint interpolation is also helpful for Light field video stabilization [60]. Recently, a deep learning based method is proposed for light field image super-resolution [61].

Denoising The reconstructed multi-view images often have high noise level which is caused by small size of the sub-aperture and aliasing. Conventional denoising methods can apply to individual viewpoint and improve the image quality, but those methods do not take care of the entire light field structure. Dansereau et al. [62, 63] developed a light field filter which can denoise the whole light field. Their approach is linear and featureless, and it performs efficiently as a single-step, constant runtime filter. An example of light field denoising is shown in Fig. 15.

5 Computer Vision and Artificial Intelligence Applications

Light field imaging has been used in many computer graphics applications for a long time. Since the light field vision has many advantages, it is getting popular to use light field imaging for computer vision and artificial intelligence applications. In this section, we mainly review the applications of 3D shape reconstruction and object detection and recognition.

5.1 3D Shape Reconstruction

A number of researches have been conducted to estimate the depth information from a light field image [12, 53, 64–69]. Wanner and Goldluecke [12] proposed method that can label the depth level globally consistent by using a variational global optimization framework. Tao et al. [64, 66] utilized shading, defocus, and correspondence in a light field image to estimate the depth information. They also developed an iterative approach to estimate and remove the specular component in order to improve the depth estimation [65]. Jeon et al. [67] proposed an approach to estimate the depth map by iterative multi-label optimization. This method can correct the distortion in lenslet image in the mean while. Lin et al. [53] took use of focal stack symmetry to recover the depth. Wang et al. [68] proposed depth estimation method with the capability to detect the occlusion area. Williem and Kyu Park [69] developed a robust depth estimation algorithm that woks well for the light field images with both occlusion and noise. SVBRDF-Invariant [70] presented a spatially-varying BRDF-invariant theory for recovering 3D shape and reflectance from light field images. Practically, they use a polynomial shape prior to resolve the ambiguity when recovering the 3D shape, and then reflectance of the object can be solved as well. Heber and Pock [71] proposed a model for recovery the shape by low rank minimization, and recently, they utilized Convolutional Neural Networks to predict depth information [72].

5.2 *Object Detection and Recognition*

There are less work for object detection and recognition using light field data, since such kind of research usually need large amount of data for training. With the availability of the commercial light field acquisition devices, more and more work has taken advantage of the light field images to deal with the challenging tasks that regular single viewpoint image can not deal with.

Saliency detection for light field images fist proposed by Li et al. [13]. The refocusing capability of light field imaging can provide useful cues, and their results show that saliency detection on light field can handle challenging scenarios such as similar foreground and background, cluttered background, complex occlusions. Zhang et al. [73] improved saliency detection accuracy by deeply investigate the depth and focus information on light fields. A comparative study of 4D light field saliency and regular 2D saliency can be found in [74].

Shimada et al. used the light field imaging for video surveillance application [11, 75]. They proposed a new feature called Local Ray Pattern (LRP) which is used to evaluate the spatial consistency of light rays. The LRP feature and GMM-based background modeling are combined to detect objects on the selected in-focus plane.

Light field imaging is also used for change detection [76, 77]. Shimada et al. [76] defined an active surveillance field (ASF) to determine in-focus and out-focus areas in the light field images, and capture temporal changes of the light rays in the ASF. Dansereau et al. [77] derived a simple but efficient solution for change detection based on closed-form method camera motion estimation from plenoptic flow [78].

Transparent objects are usually an exemption in object recognition applications. Xu et al. [14, 79] utilized the light field images to segment and classify the transparent objects. They proposed a light field distortion feature in [79], and used bag-of-feature method with the proposed feature to classify the transparent objects. Transparent object segmentation [14] can be solved by pixel labeling optimization with the constraint from light field distortion and occlusion detection.

6 Discussion and Conclusion

In this chapter, we introduce the light field vision that can solve challenging vision problems, and it can apply to many challenging AI tasks where conventional single view based vision cannot deal with. This new technique suggests several directions for new research. In this section, we give some possible directions are worthy to explore.

Light field vision for non-Lambertian object recognition and reconstruction. Light field vision has already applied to transparent object classification [79] and segmentation [14], and it will be more interesting if the target objects are generalized to all the non-Lambertian object including the specular and glossy objects. Moreover, because the light field camera captures rich 4D information of the scene, the 3D object surface can be reconstructed from the a single-shot light field image. Object recognition can be performed for the reconstructed 3D object which will make the recognition more robust.

Machine learning for light field vision. Bag-of-Feature method has been used for transparent object recognition [79], and deep learning [72] has been used for 3D shape recovery. Other machine learning methods can also apply to the light field vision. The features extracted from the light field images have more information than the features from a single-view image. Machine learning methods can utilize the features from light field images to train a sophisticated model in order to deal with different types of applications.

References

1. D. Crevier, *AI: The Tumultuous History of the Search for Artificial Intelligence* (Basic Books, 1993)
2. http://www.mobvis.org/index.htm
3. http://www.nvidia.com/object/drivepx.html
4. E.H. Adelson, J.R. Bergen, *The Plenoptic Function and the Elements of Early Vision* (Vision and Modeling Group, Media Laboratory, Massachusetts Institute of Technology, 1991)
5. T. Georgiev, A. Lumsdaine, Focused plenoptic camera and rendering. J. Electron. Imaging **19**(2), 021 106–021 106 (2010)
6. M. Levoy, P. Hanrahan, Light field rendering, in *Proceedings of the ACM Conference on Computer Graphics* (1996), pp. 31–42

7. I. Ihm, S. Park, R.K. Lee, Rendering of spherical light fields, in *Proceedings of the Fifth Pacific Conference On Computer Graphics And Applications*, ser. Pacific Graphics '97 (1997), pp. 59–68
8. http://www.viewplus.co.jp/product/camera/profusion25.html
9. http://www.lytro.com/
10. http://www.raytrix.de/
11. A. Shimada, H. Nagahara, R. ichiro Taniguchi, Object detection based on spatio-temporal light field sensing. IPSJ Trans. Comput. Vis. Appl. **5**, 129–133 (2013)
12. S. Wanner, B. Goldluecke, Globally consistent depth labeling of 4d light fields, in *IEEE Conference on Computer Vision and Pattern Recognition (CVPR)* (IEEE, 2012), pp. 41–48
13. N. Li, J. Ye, Y. Ji, H. Ling, J. Yu, Saliency detection on light field, in *IEEE Conference on Computer Vision and Pattern Recognition (CVPR)* (IEEE, 2014), pp. 2806–2813
14. Y. Xu, H. Nagahara, A. Shimada, R. Taniguchi, Transcut: transparent object segmentation from a light-field image, in *International Conference on Computer Vision (ICCV)*, 12 (2015)
15. http://cyberware.com/products/scanners/sphere.html
16. T.A. Harvey, K.S. Bostwick, S. Marschner, Measuring spatially-and directionally-varying light scattering from biological material. J. Visualized Exp. JoVE (75) (2013)
17. M. Holroyd, J. Lawrence, T. Zickler, A coaxial optical scanner for synchronous acquisition of 3d geometry and surface reflectance. ACM Trans. Graph. **29**(4), 99:1–99:12 (2010). http://doi.acm.org/10.1145/1778765.1778836
18. http://lightfield.stanford.edu/acq.html
19. A. Isaksen, L. McMillan, S.J. Gortler, Dynamically reparameterized light fields, in *Proceedings of the 27th Annual Conference on Computer Graphics and Interactive Techniques* (ACM Press/Addison-Wesley Publishing Co., 2000), pp. 297–306
20. B. Wilburn, N. Joshi, V. Vaish, E.-V.E. Talvala, E. Antunez, A. Barth, A. Adams, M. Levoy, M. Horowitz, High performance imaging using large camera arrays. ACM Trans. Graph. **24**(3), 765–776 (2005)
21. K. Venkataraman, D. Lelescu, J. Duparré, A. McMahon, G. Molina, P. Chatterjee, R. Mullis, S. Nayar, Picam: an ultra-thin high performance monolithic camera array. ACM Trans. Graph. **32**(6), 166:1–166:13 (2013). http://doi.acm.org/10.1145/2508363.2508390
22. C.-K. Liang, G. Liu, H.H. Chen, Light field acquisition using programmable aperture camera, in *IEEE International Conference on Image Processing*, vol. 5 (IEEE, 2007), pp. V–233
23. H. Nagahara, C. Zhou, T. Watanabe, H. Ishiguro, S.K. Nayar, Programmable aperture camera using lcos, in *Computer Vision-ECCV 2010* (Springer, 2010), pp. 337–350
24. C.-K. Liang, T.-H. Lin, B.-Y. Wong, C. Liu, H.H. Chen, Programmable aperture photography: multiplexed light field acquisition. ACM Trans. Graph. (TOG) **27**(3), 55 (2008)
25. A. Veeraraghavan, R. Raskar, A. Agrawal, A. Mohan, J. Tumblin, Dappled photography: mask enhanced cameras for heterodyned light fields and coded aperture refocusing. ACM Trans. Graph. **26**(3), 69 (2007). http://doi.acm.org/10.1145/1276377.1276463
26. R. Ng, M. Levoy, M. Brédif, G. Duval, M. Horowitz, P. Hanrahan, Light field photography with a hand-held plenoptic camera. Comput. Sci. Tech. Rep. CSTR (2005)
27. A. Lumsdaine, T. Georgiev, The focused plenoptic camera, in *IEEE International Conference on Computational Photography (ICCP)* (2009), pp. 1–8
28. R.Y. Tsai, A versatile camera calibration technique for high-accuracy 3D machine vision metrology using off-the-shelf TV cameras and lenses. IEEE J. Robot. Autom. **3**, 323–344 (1987)
29. Z. Zhang, A flexible new technique for camera calibration. IEEE Trans. Pattern Anal. Mach. Intell. **22**(11), 1330–1334 (2000)
30. R. Horaud, G. Csurka, D. Demirdijian, Stereo calibration from rigid motions. IEEE Trans. Pattern Anal. Mach. Intell. **22**(12), 1446–1452 (2000)
31. H. Malm, A. Heyden, Stereo head calibration from a planar object, in *IEEE Conference on Computer Vision and Pattern Recognition (CVPR)*, vol. 2 (2001), pp. II–657
32. V. Vaish, B. Wilburn, N. Joshi, M. Levoy, Using plane + parallax for calibrating dense camera arrays, in *CVPR (1)* (2004), pp. 2–9

33. T. Svoboda, D. Martinec, T. Pajdla, A convenient multi-camera self-calibration for virtual environments. PRESENCE: Teleoperators Virtual Environ. **14**(4), 407–422 (2005)
34. T. Ueshiba, F. Tomita, Plane-based calibration algorithm for multi-camera systems via factorization of homography matrices, in *ICCV* (2003), pp. 966–973
35. Y. Xu, K. Maeno, H. Nagahara, R. Taniguchi, Mobile camera array calibration for light field acquisition, in *International Conference on Quality Control by Artificial Vision (QCAV)*, vol. 5 (2013), pp. 283–290
36. R. Hartley, A. Zisserman, *Multiple View Geometry in Computer Vision* (Cambridge University Press, Cambridge, 2004)
37. N. Snavely, S.M. Seitz, R. Szeliski, Modeling the world from internet photo collections. Int. J. Comput. Vis. **80**, 189–210 (2008)
38. D.G. Dansereau, O. Pizarro, S.B. Williams, Decoding, calibration and rectification for lenselet-based plenoptic cameras, in *CVPR* (2013), pp. 1027–1034
39. D. Cho, M. Lee, S. Kim, Y.-W. Tai, Modeling the calibration pipeline of the lytro camera for high quality light-field image reconstruction, in *ICCV* (2013)
40. O. Johannsen, C. Heinze, B. Goldluecke, C. Perwaß, On the calibration of focused plenoptic cameras, in *Time-of-Flight and Depth Imaging. Sensors, Algorithms, and Applications* (Springer, 2013), pp. 302–317
41. N. Zeller, F. Quint, U. Stilla, Calibration and accuracy analysis of a focused plenoptic camera. ISPRS Ann. Photogrammetry, Remote Sens. Spat. Inf. Sci. **2**(3), 205 (2014)
42. Y. Bok, H.-G. Jeon, I.S. Kweon, Geometric calibration of micro-lens-based light-field cameras using line features, in *Proceedings of European Conference on Computer Vision (ECCV)* (2014)
43. K.H. Strobl, M. Lingenauber, Stepwise calibration of focused plenoptic cameras, Comput. Vis. Image Underst. **145**(C), 140–147 (2016). http://dx.doi.org/10.1016/j.cviu.2015.12.010
44. N. Joshi, Color calibration for arrays of inexpensive image sensors. Tech. Rep. CSTR 2004–02 (2004)
45. A. Ilie, G. Welch, Ensuring color consistency across multiple cameras, in *IEEE International Conference on Computer Vision (ICCV'05)*, vol. 2 (IEEE, 2005), pp. 1268–1275
46. Y. Chen, C. Cai, J. Liu, Yuv correction for multi-view video compression, in *18th International Conference on Pattern Recognition (ICPR'06)*, vol. 3 (IEEE, 2006), pp. 734–737
47. S.A. Fezza, M.-C. Larabi, K.M. Faraoun, Feature-based color correction of multiview video for coding and rendering enhancement. IEEE Trans. Circuits Syst. Video Technol. **24**(9), 1486–1498 (2014)
48. D.G. Lowe, Distinctive image features from scale-invariant keypoints. Int. J. Comput. Vis. **60**(2), 91–110 (2004)
49. U. Fecker, M. Barkowsky, A. Kaup, Histogram-based prefiltering for luminance and chrominance compensation of multiview video. IEEE Trans. Circuits Syst. Video Technol. **18**(9), 1258–1267 (2008)
50. S.A. Fezza, M.-C. Larabi, Color correction for stereo and multi-view coding, in *Color Image and Video Enhancement* (Springer, 2015), pp. 291–314
51. R. Ng, Fourier slice photography. ACM Trans. Graph. (TOG) **24**(3), 735–744 (2005)
52. V. Vaish, M. Levoy, R. Szeliski, C.L. Zitnick, S.B. Kang, Reconstructing occluded surfaces using synthetic apertures: Stereo, focus and robust measures, in *CVPR* (2006), pp. 2331–2338
53. H. Lin, C. Chen, S. Bing Kang, J. Yu, Depth recovery from light field using focal stack symmetry, in *Proceedings of the IEEE International Conference on Computer Vision* (2015), pp. 3451–3459
54. T.E. Bishop, S. Zanetti, P. Favaro, Light field superresolution, in *IEEE International Conference on Computational Photography (ICCP)* (IEEE, 2009), pp. 1–9
55. T. Bishop, P. Favaro, The light field camera: extended depth of field, aliasing, and superresolution. IEEE Trans. Pattern Anal. Mach. Intell. **34**(5), 972–986, 5 (2012). iNSPEC Accession Number: 12617601
56. T. Georgiev, G. Chunev, A. Lumsdaine, Superresolution with the focused plenoptic camera, in *IS&T/SPIE Electronic Imaging* (International Society for Optics and Photonics, 2011), pp. 78 730X–78 730X

57. Z. Yu, J. Yu, A. Lumsdaine, T. Georgiev, An analysis of color demosaicing in plenoptic cameras, in *IEEE Conference on Computer Vision and Pattern Recognition (CVPR)* (IEEE, 2012), pp. 901–908
58. S. Wanner, B. Goldluecke, Variational light field analysis for disparity estimation and super-resolution. IEEE Trans. Pattern Anal. Mach. Intell. (2014)
59. A. Levin, F. Durand, Linear view synthesis using a dimensionality gap light field prior, in *IEEE Conference on Computer Vision and Pattern Recognition (CVPR)* (IEEE, 2010), pp. 1831–1838
60. B.M. Smith, L. Zhang, H. Jin, A. Agarwala, Light field video stabilization, in *12th international conference on computer vision* (IEEE, 2009), pp. 341–348
61. Y. Yoon, H.-G. Jeon, D. Yoo, J.-Y. Lee, I. So Kweon, Learning a deep convolutional network for light-field image super-resolution, in *Proceedings of the IEEE International Conference on Computer Vision Workshops* (2015), pp. 24–32
62. D.G. Dansereau, D.L. Bongiorno, O. Pizarro, S.B. Williams, Light field image denoising using a linear 4d frequency-hyperfan all-in-focus filter, in *Proceedings of the SPIE Conference on Computational Imaging (SPIE)*, vol. 8657 (2013)
63. D.G. Dansereau, O. Pizarro, S.B. Williams, Linear volumetric focus for light field cameras. ACM Trans. Graph. (TOG) **34**(2), 15 (2015)
64. M.W. Tao, S. Hadap, J. Malik, R. Ramamoorthi, Depth from combining defocus and correspondence using light-field cameras, in *Proceedings of the IEEE International Conference on Computer Vision* (2013), pp. 673–680
65. M.W. Tao, T.-C. Wang, J. Malik, R. Ramamoorthi, Depth estimation for glossy surfaces with light-field cameras, in *Workshop on Light Fields for Computer Vision, ECCV* (2014)
66. M.W. Tao, P.P. Srinivasan, J. Malik, S. Rusinkiewicz, R. Ramamoorthi, Depth from shading, defocus, and correspondence using light-field angular coherence, in *2015 IEEE Conference on Computer Vision and Pattern Recognition (CVPR)* (IEEE, 2015), pp. 1940–1948
67. H.-G. Jeon, J. Park, G. Choe, J. Park, Y. Bok, Y.-W. Tai, I.S. Kweon, Accurate depth map estimation from a lenslet light field camera, in *IEEE Conference on Computer Vision and Pattern Recognition (CVPR)* (IEEE, 2015), pp. 1547–1555
68. T.-C. Wang, A.A. Efros, R. Ramamoorthi, Occlusion-aware depth estimation using light-field cameras, in *Proceedings of the IEEE International Conference on Computer Vision* (2015), pp. 3487–3495
69. W. Williem, I. Kyu Park, Robust light field depth estimation for noisy scene with occlusion, in *Proceedings of the IEEE Conference on Computer Vision and Pattern Recognition* (2016), pp. 4396–4404
70. T.-C. Wang, M. Chandraker, A. Efros, R. Ramamoorthi, Svbrdf-invariant shape and reflectance estimation from light-field cameras, in *Proceedings of the IEEE Conference on Computer Vision and Pattern Recognition (CVPR)* (2016)
71. S. Heber, T. Pock, Shape from light field meets robust pca, in *European Conference on Computer Vision* (Springer, 2014), pp. 751–767
72. S. Heber, T. Pock, Convolutional networks for shape from light field, in *Proceedings of the IEEE Conference on Computer Vision and Pattern Recognition* (2016), pp. 3746–3754
73. J. Zhang, M. Wang, J. Gao, Y. Wang, X. Zhang, X. Wu, Saliency detection with a deeper investigation of light field, in *Proceedings of the 24th International Joint Conference on Artificial Intelligence* (2015), pp. 2212–2218
74. X. Zhang, Y. Wang, J. Zhang, L. Hu, M. Wang, Light field saliency vs. 2d saliency: a comparative study. Neurocomputing **166**, 389–396 (2015)
75. A. Shimada, H. Nagahara, R.-I. Taniguchi, Background light ray modeling for change detection. J. Vis. Commun. Image Representation **38**, 55–64 (2016)
76. A. Shimada, H. Nagahara, R.-I. Taniguchi, Change detection on light field for active video surveillance, in *12th IEEE International Conference on Advanced Video and Signal Based Surveillance (AVSS)* (IEEE, 2015), pp. 1–6
77. D.G. Dansereau, S.B. Williams, P.I. Corke, Simple change detection from mobile light field cameras. Comput. Vis. Image Underst. **145**, 160–171 (2016)

78. D.G. Dansereau, I. Mahon, O. Pizarro, S.B. Williams, Plenoptic flow: closed-form visual odometry for light field cameras, in *2011 IEEE/RSJ International Conference on Intelligent Robots and Systems* (IEEE, 2011), pp. 4455–4462
79. Y. Xu, K. Maeno, H. Nagahara, A. Shimada, R. Taniguchi, Light field distortion feature for transparent object classification. Comput. Vis. Image Underst. (2015)

Author Index

H. Lu and Y. Li (eds.), *Artificial Intelligence and Computer Vision*,
Studies in Computational Intelligence 672, DOI 10.1007/978-3-319-46245-5

Printed in the United States
By Bookmasters